BRITISH GEOLOGICAL SURVEY

DTI MINERALS PROGRAMME PUBLICATI(

Exploration for Metalliferous and Related Minerals in Britain: A Guide

2nd Edition

Authors
T B Colman, MSc, MIMM and D C Cooper PhD, CGeol

The National Grid and other Ordnance Survey data are used with the permission of the Controller of Her Majesty's Stationery Office.
Ordnance Survey licence number GD 272191/2000

ISBN 0 85272 357 1

Front cover

Top Spectral image from PIMA™ mineral analyser.
Centre BGS drilling for PGE in north-east Scotland. *Photograph by Mick Strutt.*
Bottom Prospectivity analysis for gold in the Tyndrum area of Scotland.

Bibliographical reference

Colman, T B and Cooper, D C. Exploration for Metalliferous and Related Minerals in Britain: A Guide (2nd Edition). DTI Minerals Programme Publication No. 1.

Keyworth, Nottingham British Geological Survey 2000

BRITISH GEOLOGICAL SURVEY

The full range of Survey publications is available from the BGS Sales Desk at the Survey headquarters, Keyworth, Nottingham. The more popular maps and books may be purchased from BGS-approved stockists and agents and over the counter at the Bookshop, Gallery 37, Natural History Museum (Earth Galleries), Cromwell Road, London. Sales desks are also located at the BGS London Information Office, and at Murchison House, Edinburgh. The London Information Office maintains a reference collection of BGS publications including maps for consultation. Some BGS books and reports may also be obtained from the Stationery Office Publications Centre or from the Stationery Office bookshops and agents.

The Survey publishes an annual catalogue of maps, which lists published material and contains index maps for several of the BGS series.

The British Geological Survey carries out the geological survey of Great Britain and Northern Ireland (the latter as an agency service for the government of Northern Ireland), and of the surrounding continental shelf, as well as its basic research projects. It also undertakes programmes of British technical aid in geology in developing countries as arranged by the Department for International Development and other agencies.

The British Geological Survey is a component body of the Natural Environment Research Council.

This book has been prepared for the Department of Trade and Industry by the Minerals Group, British Geological Survey.

British Geological Survey offices

Keyworth, Nottingham NG12 5GG
☎ 0115 936 3100 Fax 0115 936 3200
e-mail: sales@bgs.ac.uk
www.bgs.ac.uk
www.british-geological-survey.co.uk

Murchison House, West Mains Road, Edinburgh EH9 3LA
☎ 0131 667 1000 Fax 0131 668 2683

London Information Office at the Natural History Museum, Earth Galleries, Exhibition Road, South Kensington London, SW7 2DE
☎ 0171 589 4090 Fax 0171 584 8270
☎ 0171 938 9056/57

Forde House, Park Five Business Centre, Harrier Way, Sowton Exeter EX2 7HU
☎ 01392 445271 Fax 01392 445371

Aberystwyth, Room G19, Sir George Stapledon Building, University of Wales, Penglais, Aberystwyth, Ceredigion, Wales, SY23 3DB
☎ 01970 622541 Fax 01970 622542

Geological Survey of Northern Ireland, 20 College Gardens, Belfast BT9 6BS
☎ 02890 666595 Fax 02890 662835
e-mail: gsni@bgs.ac.uk
www.bgs.ac.uk/gsni

Maclean Building, Crowmarsh Gifford, Wallingford, Oxfordshire OX10 8BB
☎ 01491 838800 Fax 01491 692345
e-mail: hydro@bgs.ac.uk

Parent Body

Natural Environment Research Council
Polaris House, North Star Avenue, Swindon, Wiltshire SN2 1EU
☎ 01793–411500 Fax 01793–411501

Contents

Appendices

Tables

Figures

Foreword

Britain is fortunate in being blessed with mineral resources, but its mineral wealth and potential is often underestimated. Although Britain has produced minerals since before Roman times, new deposits continue to be found by the application of new geological models and improved mineral exploration techniques.

This publication provides a comprehensive guide for those planning minerals exploration and development in Britain. It briefly summarises the geology, perceived mineral potential of Britain, mineral exploration techniques commonly used here and the legal framework and planning regulations that apply to mineral developments. It also provides contact details for many organisations which can be approached for further information on specific topics, and lists reports that describe previous mineral exploration work.

Prime examples of recent private-sector exploration initiatives include the assessment of recently discovered gold mineralisation in an area of Devon previously considered to have little potential, and exploration for gemstone deposits in Scotland and Northern Ireland. Several gold prospects have been found in Scotland and Northern Ireland, and proposals for mine development at two sites have received approval.

Britain has a highly developed economy, excellent infrastructure, a skilled workforce, and a government committed to encouraging business and investment in Britain within a fair and effective legal and regulatory framework. Government is also committed to promoting small business growth and making the most of the UK's science, engineering and technology base through initiatives such as the Foresight Programme. Our aim is to generate higher levels of sustainable growth and productivity in a modern economy and I invite you to help us achieve it. I therefore hope that you will find this guide helpful and wish you every success with your work here.

Alan Johnson MP
Minister for Competitiveness,
Department of Trade and Industry

Summary

Britain is, in relation to its size, remarkably well-endowed with mineral resources; the type and distribution of which are related to the wide range of geological environments represented in the British Isles and the adjacent continental shelf. Mining has been an industry here for at least 2000 years and recent discoveries demonstrate that there are still new metalliferous mineral deposits to be found. A brief review of the geology and structure of Britain, in the light of exploration activity and discoveries since 1965, leads to the conclusion that Britain has most economic potential for the following types of metalliferous mineral deposit:

- Sedimentary exhalative deposits of baryte, base-metals and gold.
- Carbonate-hosted deposits of Irish and Pennine sub-types.
- Mesothermal gold deposits, predominantly of slate-belt (turbidite-hosted) type.
- Unconformity related (redox) gold deposits.
- Volcanogenic deposits of base-metals, silver and gold, including volcanic massive sulphide deposits.
- Mafic-intrusion-related nickel-copper deposits.
- Gemstones in deep-seated minor intrusions.
- Epithermal gold deposits associated with Devonian volcanism.
- Granite-related polymetallic vein and stockwork mineralisation, principally for tin and tungsten.

Other types present with some potential include offshore placers of tin and chromite, uranium in granite-related veins and sedimentary stratabound concentrations, copper, molybdenum and gold in porphyry-style settings, and copper in sedimentary stratabound red-beds.

The rights to non-fuel minerals in Great Britain, with the exception of gold and silver, are mainly in private ownership. The mineral rights to gold and silver in most of Britain are owned by the Crown, and a licence for the exploration and development of these metals must be obtained from the Crown Estate Commissioners through the Crown Mineral Agent. The right to exploit minerals in the foreshore (beach) and on the seabed within the limits of national jurisdiction is also vested in the Crown and, apart from coal, oil and natural gas, these resources are managed by the Crown Estate Commissioners. Mineral rights for other commodities are generally held by the current surface landowner, though they may have been retained by a previous landowner, particularly in areas with a long history of mining. There is no national register of mineral rights, but the Mines Working Facilities and Support Act 1966, as amended, provides a means by which an operator who is either unable to trace the mineral rights owner, or cannot reach an agreement on reasonable terms with the owner, can obtain the authority to explore for and work minerals. All minerals in Northern Ireland, except gold and silver (already owned by the Crown) and 'common' substances including sand and gravel and aggregates, are vested in the Department of Enterprise, Trade and Industry (DETI) which grants prospecting and mining licences.

Most land is owned by the occupier or farmer who holds the surface rights. Permission must generally be obtained from the surface landowner to gain access to land for mineral exploration purposes. The time and effort required to establish prospecting rights is very variable, depending in part on the size of landholdings in the area. The approach of companies varies, but many prefer to have a flexible legal agreement allowing access for surface prospecting, with the right of a first refusal to an option over the mineral rights after a specified time.

As in other developed countries, Britain is subject to planning controls governing most forms of land development. Day-to-day responsibility for administering the planning system rests with the local planning authorities (LPAs). The key feature of the planning system is that most forms of development require planning permission before development can take place. Certain operations, including most short-term mineral exploration activities that have minimal environmental impact, do not require specific planning permission. However, companies are strongly advised to discuss their plans with the LPA's Minerals Officer, who will be able to advise them on what information will be required to help the planning committee reach a decision in a particular case. Each application is considered on its merits and will balance perceived benefits and detriment. If the application is refused, or the conditions are unacceptable, the applicant has the right of appeal to the Secretary of State for the Department of the Environment, Transport and the Regions (DETR). An appeal following refusal of a minerals application usually results in a public inquiry, conducted by a planning inspector. The inspector has delegated powers to determine the appeal. However, in certain circumstances, the Secretary of State will make the decision, taking account of the inspector's recommendations. In recent years planning permission has been granted for two gold mines but refused for a baryte mine.

In Northern Ireland planning permission is not usually required for normal prospecting activities. However, the Department of the Environment (DoE) in Northern Ireland, which is responsible for planning, is consulted by the DETI before a prospecting licence is issued. Licensees are required to notify the DoE of the nature and scale of their proposed activities.

In Britain the government department with responsibility for metalliferous, industrial and energy minerals is the Department of Trade and Industry (DTI). Construction (bulk) minerals fall within the remit of the DETR. The British Geological Survey (BGS) is the main Government agency in Great Britain for undertaking national surveys in the earth sciences and is the recognised national repository for geoscience data. The Geological Survey of Northern Ireland (GSNI) has a similar function but is an office of the DETI. Information and advice on mineral exploration and development in Britain is available from BGS and GSNI. Principal datasets held by the BGS relevant to mineral exploration include:

- Geological mapping at various scales from 1:10,000 to 1:1 m.
- Regional and local scale geochemical survey data and atlases.

- Regional and local scale ground and airborne geophysical data.
- Open-file company reports provided under the terms of the Mineral Exploration and Investment Grants Act 1972.
- Public sector (BGS Mineral Reconnaissance Programme) reports and data releases.
- Mineral occurrence and mineral workings databases.
- Drillcore, rock samples and thin sections.
- World-wide mineral trade and productions statistics.

The UK offers one of the best business locations in the world. It has a highly developed infrastructure and successive British Governments have actively supported and encouraged inward investment and new business developments. It does not discriminate against foreign companies, constrain repatriation of profits or limit borrowing. It also offers fast, easy access to and within the European Union, a skilled and adaptable workforce and long-term stability. As a result, it is home to some of the world's greatest companies and is regarded by them as the business centre for Europe.

Information related to business development is provided by the DTI. The Invest in Britain Bureau (IBB), a unit within the DTI, is the Government's main agency for promoting inward investment. The Bureau operates through British Embassies, High Commissions and Consulates. Within Britain it works through a network of offices in major cities. The IBB can assist firms with all aspects of locating or relocating a business in the UK or expanding existing facilities. It is the central contact point in Britain for all advice and assistance.

Other useful information sources include the Crown Mineral Agent, Ordnance Survey, National Remote Sensing Centre, British Library, professional institutions and minerals industry representative organisations. First points of contact for further information are:

Onshore Minerals and
Energy Resources Programme
British Geological Survey Keyworth
NOTTINGHAM NG12 5GG UK
Tel +44(0)115 9363494
Fax +44(0)115 936 3520
E-mail: minerals@bgs.ac.uk
Internet www.mineralsuk.com

Department of Trade and Industry
Metals, Minerals and Shipbuilding
Directorate
151 Buckingham Palace Road
LONDON SW1W 9SS UK
Tel +44(0)207 215 1102
Fax +44(0)207 215 1070
Internet www.dti.gov.uk

Geological Survey of Northern Ireland
20 College Gardens
BELFAST BT9 6BS
Northern Ireland
Tel +44(0)2890 666595
Fax +44(0)2890 662835
E-mail: gsni@bgs.ac.uk
Internet www.bgs.ac.uk/gsni

1 INTRODUCTION

Britain comprises Great Britain (England, Scotland and Wales) and Northern Ireland; its full name is the United Kingdom (UK) of Great Britain and Northern Ireland (The Stationary Office, 1999). Britain constitutes the greater part of the British Isles, a geographical term for the group of islands off the north-west coast of Europe. The Isle of Man in the Irish Sea and the Channel Islands between Great Britain and France are largely self-governing; they are not part of the United Kingdom.

The UK offers one of the best business locations in the world. It has a good deal to offer — fast, easy access to and within the European Union, a skilled and adaptable workforce and a business-friendly environment. It offers stability, enabling companies to plan and invest for the long term. The UK is already home to some of the world's greatest companies and regarded by them as the business centre for Europe.

Britain is, in relation to its size, remarkably well-endowed with mineral resources, the type and distribution of which are related to the complex geological and tectonic history of the British Isles and the adjacent continental shelf. Non-ferrous metal mining has been an industry of major importance in several regions. In south-west England tin, copper and other metals have been mined for at least 2000 years from deposits related to Variscan granites; in the Pennine hills of central and northern England, lead, and lately fluorite and baryte, have been extracted from vein and replacement deposits in Carboniferous sediments over a similar period; in Central Wales, the Isle of Man and the Southern Uplands of Scotland lead and zinc have been mined from vein deposits in Lower Palaeozoic greywackes, and in the Lake District copper, lead and zinc have been recovered from a variety of deposits in Lower Palaeozoic sedimentary and volcanic rocks underlain by Caledonian granites. Gold occurs principally in Scotland, Wales and south-west England. The formerly important Jurassic ironstones of central and eastern England, the Carboniferous hematites of the Lake District and South Wales and the ironstones of the Coal Measures have been almost entirely abandoned due to competition from high-grade imported ores and the exhaustion of some deposits.

Crude oil and natural gas have become the dominant mineral commodities by value during the last three decades with the development of the North Sea as a major petroleum province. The importance of coal mining to the national economy has declined considerably in the last decade but there were still 75 deep (including shallow drift) mines and 118 opencast sites working in 1998. Major discoveries of Tertiary lignite have been made in Northern Ireland.

Britain is also a major producer of constructional and industrial minerals, including barytes, ball clay, china clay, fluorspar, fuller's earth, gypsum, potash, salt and silica sand. Annual production of minerals is shown in Table 1.

2 THE PUBLIC SECTOR ROLE IN MINERAL EXPLORATION

The Department of Trade and Industry (DTI) has responsibility within Government for the metalliferous and industrial minerals mining industry. The DTI aims to help business to compete successfully at home, in the rest of Europe and throughout the world. It has a wide range of overall responsibilities including industrial sponsorship, export promotion, inward investment, energy policy, science and technology, investor protection, and support for small firms. The DTI supports the BGS Minerals Programme (see below) and publishes information leaflets and guides for businesses.

The Invest in Britain Bureau (IBB), jointly funded by the DTI and the Foreign Office, is the main Government agency for inward investment and promotes the whole of the UK as an investment location. The IBB can assist firms with all aspects of locating or relocating a business in the UK or expanding existing facilities. It is the central contact point in Britain for all advice and assistance. The Bureau operates overseas through British Embassies, High Commissions and Consulates-General. Within Britain it works through a network of offices in major cities. Further details of contact points are given in Section 8.

Between 1972 and 1984 the DTI provided financial assistance to industry for mineral exploration in Great Britain through the Mineral Exploration and Investment Grants Act 1972 (MEIGA). The Act required that information from prospecting work carried out under its regulations should be deposited with the BGS and, as a result, data on more than 150 projects are now available on open file. Open file data from the MEIGA programme are referred to in the text by the prefix MEG and the project number (for example MEG 212).

The Department of the Environment, Transport and the Regions (DETR) has responsibility within Government for bulk and construction minerals. Its other responsibilities include local government, regional development agencies the development and implementation of the planning regulations, countryside affairs, environmental protection and water. The DETR funds the UK Minerals Yearbook produced by the BGS, and places contracts with the BGS and private-sector companies to undertake nationwide studies related to the availability of minerals and the relationship between their occurrence and land use to aid planning policy. Planning legislation which affects mineral exploration is covered in Section 7.

In Scotland, the Scottish Office, headed by the Secretary of State for Scotland, is responsible for a wide range of statutory functions which in England and Wales are the responsibility of a number of departmental ministers. These include industry, planning and development. Similarly, the Welsh Office has responsibility in Wales for many of the same ministerial functions, including town and country planning, land use, water, and financial assistance to industry. Now that the devolution acts have been implemented both Scotland and Wales have their own parliaments and ministers who have responsibility for these matters.

In Northern Ireland power was devolved to the Northern Ireland Assembly and its Executive Committee of Ministers in December 1999. The Northern Ireland Assembly is responsible for the government of the Province. All mineral exploration and development in Northern Ireland is carried out under licence from the Department of Enterprise, Trade and Investment (DETI) — formerly the Department of Economic Development (DED) whose remit also includes responsibility for promotion of inward investment, business development, health and safety, energy, company regulation, and industrial relations. The DETI publishes a triennial report on Mineral Exploration and Development in Northern Ireland.

The British Geological Survey (BGS) is the main Government agency in Great Britain for undertaking national work in the earth sciences and is the recognised national repository for geoscience data. The BGS is a component body of the Natural Environment Research Council (NERC). The principal objectives of the BGS are to:

Table 1 United Kingdom production of minerals 1991–1998.

Thousand tonnes

Mineral	1991	1992	1993	1994	1995	1996	1997	1998 (Estimated)
Coal:								
Deep-mined	73 357	65 800	50 457	31 854	35 150	32 223	30 281	**25 085**
Opencast	18 636	18 187	17 006	16 804	16 369	16 315	16 700	**14 824**
Other (a)	2 209	507	736	313	1 518	1 658	1 514	**1 364**
Natural gas and oil:								
Methane (oil equivalent)								
Colliery	86	78	69	63	52	49	45	**...**
Onshore	147	190	212	226	304	84 127	85 805	**90 073**
Offshore	50 378	51 197	60 223	64 308	70 407			
Other (b)	255	306	319	359	379	442	500	**...**
Crude oil								
Onshore	3 703	3 962	3 737	4 649	5 051	5 251	4 981	**124 220**
Offshore	83 129	85 222	90 213	114 383	116 743	116 679	115 135	
Condensates and other (c)								
Onshore	153	171	157	210	221	8 077	8 089	**8 521**
Offshore	4 275	4 896	6 082	7 697	8 309			
Iron ore	59	31	1.1	1.3	1.1	1.2	1.2	**1.2**
Non-ferrous ores (metal content):								
Tin	2.3	2.0	2.2	1.9	2.0	2.1	2.4	**0.4**
Lead	(i) 1.0	(i) 1.0	(i) 1.0	2	(i) 1.6	(i) 1.8	(i) 1.6	**1.6**
Copper (d)	0.3	—	—	—	—	—	—	
Zinc (d)	1.1	...	...	...	...	...	—	**—**
Silver (e) (kg)	565	—	—	—	—	—	...	
Gold (kg)	...	...	...	...	...	...	...	**...**
Chalk (p)	10 317	9 171	9 076	10 236	9 949	9 239	9 550	**9 500**
Common clay and shale (p)	13 050	12 155	10 891	12 464	13 930	11 804	11 322	**11 500**
Igneous rock (k) (l)	53 948	57 654	57 766	56 494	57 205	50 705	48 656	**49 000**
Limestone (excluding dolomite)	94 861	89 399	93 727	106 626	94 636	86 564	87 752	**106 000**
Dolomite (excluding limestone)	19 454	18 539	17 985	17 616	17 966	16 555	17 282	
Sand and gravel:								
Land	89 311	82 037	83 698	91 450	83 293	78 173	79 500	**100 000**
Marine (j)	17 053	16 874	16 319	17 969	18 439	18 204	18 883	
Sandstone	16 607	14 890	16 059	18 974	19 796	17 522	18 499	**18 700**
Slate (h)	360	326	462	402	275	408	347	**450**
Ball clay (sales)	729	744	746	825	893	880	916	**964**
Barytes	86	77	(i) 55	(i) 54	(i) 85	93	(i) 74	**68**
Calcspar	8	4	3	...	...	...	13	**...**
Celestite	2	2	2	—	—	—	—	**—**
Chert and flint (f)	5	...	...	...	...	...	...	**...**
China clay (sales) (n)	2 911	2 502	2 461	2 530	2 586	2 281	2 360	**2 400**
China stone	6	8	7	8	9	8	8	**8**
Diatomite (n)	0.2	0.1	0.2	0.2	—	—	—	**—**
Fireclay (p)	867	572	479	679	708	536	338	**(e) 800**
Fluorspar	78	76	70	(i) 58	(i) 55	(i) 65	(i) 64	**65**
Fuller's earth (sales) (g)	189	189	187	134	132	143	135	**94**
Gypsum (natural) (i)	2 500	2 500	2 500	2 000	2 000	2 000	2 000	**2 000**
Lignite	3	3	3	2	—	...	...	**...**
Peat (000 m^3)	1 561	1 506	1 452	1 982	2 280	1 885	1 619	**1 500**
Potash (o)	825	883	941	966	933	1 030	941	**1 014**
Rock salt (i)	1 635	1 400	1 200	1 700	1 800	2 200	1 800	**700**
Salt from brine (i)	1 320	1 300	1 300	1 300	1 300	1 300	1 300	**1 300**
Salt in brine (m)	3 874	3 401	4 076	4 009	3 548	3 512	3 561	**3 500**
Silica sands	4 201	3 615	3 587	4 038	4 344	4 861	4 704	**4 600**
Talc	11	5	5	5	4	5	6	**5**

(a) Slurry etc. recovered from dumps, ponds, rivers etc.
(b) Biogas from landfill and sewage.
(c) Including ethane, propane and butane, in addition to condensates.
(d) Content of mixed concentrate.
(e) Previous figures are believed to be too low.
(f) Great Britain only.
(g) BGS estimates based on data from producing companies.
(h) Slate figures include waste used for constructional fill and powder and granules used in industry.
(i) BGS estimate.
(j) Including marine-dredged landings at foreign ports (exports); see p.5–16.
(k) Excluding a small production of granite in Northern Ireland.
(l) In addition, the following amounts of igneous rock were produced in Guernsey (thousand tonnes): 1992: 151; 1993: 180; 1994: 192; 1995: 184; 1996: 198; 1997: 115, and Jersey: 1996: 348; 1997: 350.
(m) Used for purposes other than salt making.
(n) Dry weight.
(o) Marketable product (KCl).
(p) Excluding a small production in Northern Ireland.

Source: United Kingdom Minerals Yearbook 1998. (Keyworth, Nottingham: British Geological Survey, 1999)

- Advance geoscientific knowledge of the United Kingdom landmass and its adjacent continental shelf by means of systematic surveying and data collection, long-term monitoring and high quality research.
- Provide comprehensive, objective, impartial and up to date geoscientific information, advice and services which meet the needs of customers in the industrial, engineering, governmental and scientific communities of the UK and overseas, thereby contributing to the economic competitiveness of the United Kingdom, the effectiveness of public services and policy, and quality of life.
- Enhance the UK science base by providing knowledge, information, education and training in the geosciences, and promote the public understanding of the relevance of geoscience to resource and environmental issues.

A major task of the BGS is to prepare, and keep under revision, the geological synthesis of the landmass of Great Britain and the adjacent continental shelf, and to publish the results as maps (primarily at 1:10 000, 1:50 000 and 1:250 000 scales) and reports. The BGS also carries out the Geochemical Baseline Survey of the Environment (G-BASE), nationwide ground and airborne geophysical surveys, and the DTI-funded Minerals Programme, all of which provide base-line data of value for mineral exploration. The BGS is now able to provide much of its minerals-related information in digital form using its MINGOL (Minerals GIS On-Line) information system (Section 8.1.8).

The Geological Survey of Northern Ireland (GSNI) has a function similar to that of the BGS but is an office of the Department of Enterprise, Trade and Industry (DETI). In Northern Ireland, in contrast to Great Britain, the ownership of most minerals is vested in the State (Section 7) and the DETI is responsible for granting exploration and mining licences. The GSNI, which holds a large amount of data on open file, advises the Department and assists in administering licences.

Further information, including contact addresses for the above organisations and products, are given in data sources (Section 8).

3 GEOLOGY AND STRUCTURE

The geology of Britain is extremely varied and complex, but it can be divided into a relatively small number of chronostratigraphic units (Figure 1) and a generalised tectonic framework (Figure 2). The BGS Tectonic Map (British Geological Survey, 1996c) provides a detailed compilation of the structure of the British Isles. The following summary is based on a large volume of published information, but notable sources include Anderton et al. (1979), Craig (1991), Duff and Smith (1992), Hancock (1983), Harris et al. (1979) and Windley (1984). The Atlas of Palaeogeography and Lithofacies (Cope et al., 1992) is an excellent overview of the geological development of Britain. More detailed information is given in the series of handbooks on the regional geology of Great Britain and Northern Ireland published by HMSO for the British Geological Survey.

3.1 Precambrian

The oldest rocks are the Precambrian (Archaean) Lewisian high-grade gneisses (2900 Ma) of north-west Scotland. They include fragmented layered basic and ultrabasic intrusions and also younger Lower Proterozoic (2000 Ma) supracrustal sedimentary and volcanic rocks, such as those which host the volcanogenic Gairloch Cu-Zn deposit. The rocks lie on the eastern margin of the North Atlantic craton and form the North-west Foreland (Figures 2 and 3). They occur mainly west of the Moine Thrust and are unconformably overlain by thick, undeformed Proterozoic (1000 Ma) Torridonian red-bed arkoses and siltstones which are almost entirely unmineralised. The thick, shallow water clastic sediments of the Proterozoic (1250 Ma) Moine Supergroup lying to the east of the Moine Thrust are highly tectonised and metamorphosed to granulite facies in places. They are practically unmineralised except where they contain tectonic slices of the underlying Lewisian basement. Another Precambrian block, formed by the Midlands Microcraton of Central England (Figure 2), is almost completely concealed by younger rocks but there are small outcrops of late Precambrian volcano-sedimentary sequences in the Welsh borders and the English Midlands. Its boundaries are largely conjectural (Pharaoh et al., 1987). Late Precambrian rocks also occur in Anglesey, North Wales.

3.2 Dalradian

The Moine Supergroup is overlain to the south-east by the more lithologically diverse late Proterozoic to early Palaeozoic Dalradian Supergroup. This succession contains thick clastic sediments with local basic volcanics, deposited in a series of extensional ensialic basins that formed on the north side of the expanding Iapetus Ocean which opened between the North Atlantic and Eurasian continents in late Precambrian times. The world-class Aberfeldy Ba + (Pb-Zn) sedimentary exhalative (SEDEX) deposits occur in the Middle Dalradian of Central Scotland, while a minor volcanogenic Cu-Zn deposit has been found at Vidlin in Shetland. Gold mineralisation has been found at several localities. Accretion of the Precambrian Southern Uplands basement onto the north-west Foreland during the early Ordovician caused extensive tectonism in the Moine and Dalradian successions, the Grampian orogeny, with metamorphism reaching upper amphibolite grade. Major intrusive basic and ultrabasic bodies, including layered intrusions with Cu-Ni and PGE potential, were emplaced into the Dalradian succession in north-east Scotland during the Ordovician. They were subsequently deformed and dismembered during the Caledonian orogeny.

3.3 Lower Palaeozoic

Subduction of the northern Iapetus Ocean floor switched in Ordovician times from under the Midland Valley to under the Southern Uplands basement and an island arc developed. The Ballantrae ophiolite sequence was obducted onto the basement margin. Lower Palaeozoic formations in Wales and the Lake District consist of basic to acid volcanics and volcaniclastic sediments with thick basinal greywacke turbidites and co-eval shelf sediments. The two continental blocks of the North Atlantic and Eurasian cratons were on opposite sides of the Iapetus Ocean, which was gradually closing in a complex way with subduction on both sides. The Southern Uplands of Scotland, together with the Down-Longford massif in Northern Ireland, formed an accretionary back arc and foreland basin complex on the leading edge of the North Atlantic block. Extensive volcanism occurred during the Ordovician in the Lake District and Wales (and also in the concealed Lower Palaeozoic rocks to the east of the Midlands Microcraton) as southward subduction of the Iapetus Ocean floor continued. Thin Cambrian quartzites and limestones were deposited on the continental shelf of the North-west Foreland and the Midlands Microcraton.

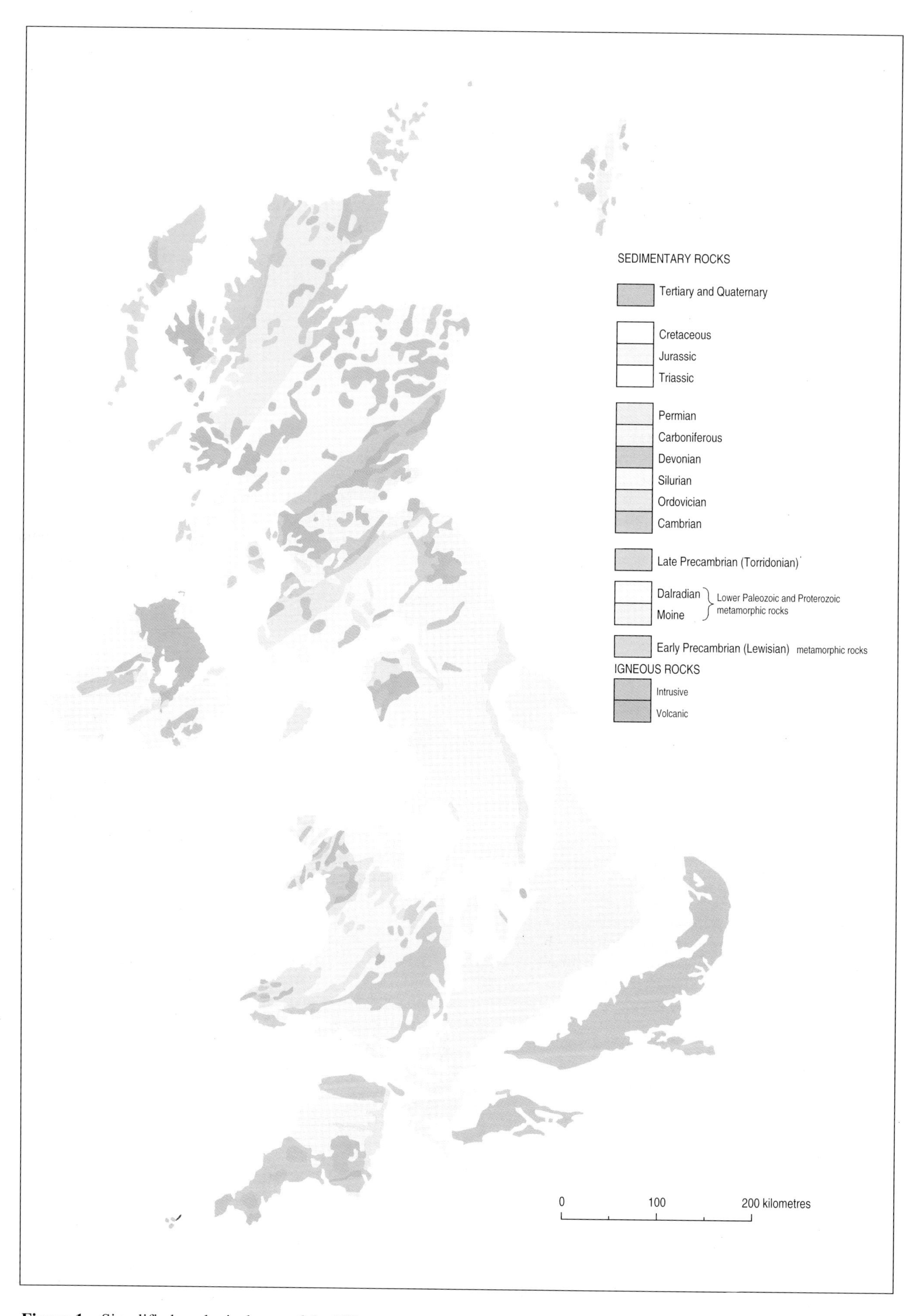

Figure 1 Simplified geological map of the UK.

Mineralisation occurred in North Wales during the Lower Palaeozoic, with the formation of the Coed-y-Brenin porphyry copper deposit in Cambrian diorite, the Parys Mountain Zn-Cu-Pb volcanogenic massive sulphide deposit in Ordovician/Silurian volcanic and sedimentary rocks, stratbound iron and manganese deposits and widespread minor Cu-Pb-Zn vein mineralisation in Snowdonia.

3.4 Caledonian orogeny

Closure of the Iapetus Ocean was complete by late Silurian to early Devonian times along the line of the Iapetus suture in northern England (Figure 2). There was further deformation of the Moine and Dalradian rocks on the north side of the suture, accompanied by metamorphism up to granulite grade and widespread granitic intrusion. The Southern Uplands accretionary complex, and rocks to the south side of the Iapetus Suture, were less highly tectonised. The main effect was the development of a pervasive slaty cleavage. Exposed granites are less common than in the Moine and Dalradian areas, but a number of concealed granites have been indicated by gravity studies in the last thirty years. Some, such as those underlying Weardale and Wensleydale in northern England, have been proved by drilling. Others, such as those around the Wash in eastern England, remain unproven. Lower Palaeozoic rocks overlying the Midlands Microcraton were almost undeformed.

The end-Caledonian orogeny is associated with a period of intense metalliferous mineralisation. Granite-associated Cu, Au, Mo and W mineralisation occurs in Scotland and the Lake District. Sediment-hosted Pb-Zn ± Cu ± Ba vein mineralisation occurs in Central Wales, Shropshire, the Southern Uplands, the Lake District, Shropshire and the Isle of Man. Turbidite-hosted gold mineralisation occurs in Wales, the Lake District and southern Scotland. Quartz veins of probable Caledonian age, with economic Au-Ag mineralisation, occur in the Dalradian of Scotland and Northern Ireland.

3.5 Upper Palaeozoic

The Upper Palaeozoic Devonian to Permian successions were mainly deposited on a stable platform and consist predominantly of clastic and carbonate sediments. Local extensional basins, such as those of Craven and Solway-Northumberland, developed in the early Carboniferous. General continued subsidence led to the deposition of thick deltaic mudstones and sandstones with associated workable coals in the late Carboniferous (Westphalian). These were followed in Permian times by red-beds containing thick evaporites. The Midland Valley of Scotland developed as an internal molasse trough during the early Devonian, with thick calc-alkaline lavas and clastic red-bed facies sediments derived from the erosion of the adjacent Caledonian mountains. It is bounded to the north by the Highland Boundary Fault and to the south by the Southern Uplands Fault.

In south-west England block and basin limestones and shales, spilitic basic volcanics and turbidite deposits were laid down in an extensional basin developed at the western end of the Rheno-Hercynian zone. Shallow-water clastic sediments were deposited in north Devon, which lay on the northern margin of the basin.

Contemporaneous mineralisation includes uranium in the Devonian lacustrine succession of the Orcadian basin in northern Scotland, and salt, gypsum and potash deposits of Upper Permian age in northern England. Minor synsedimentary Zn-Pb mineralisation occurs in the early Carboniferous of the Craven Basin and around the Devonian/Carboniferous boundary in south-west England. The Craven Basin mineralisation shows some similarities to the important early Carboniferous Zn-Pb deposits in the Republic of Ireland, such as those at Navan. Disseminated and vein-style gold mineralisation has recently been discovered close to the junction of Permian red-beds and associated alkaline basalt lavas with older rocks in south-west England and southern Scotland.

3.6 Variscan orogeny

The Variscan orogeny from late Devonian to early Permian times was caused by a complex series of movements and collisions between Europe, Africa and North America. The main areas affected in Britain were the Variscides in southern England and South Wales (Figure 2). The orogeny caused deformation and tectonism and culminated in the emplacement of the high-heat-flow, Cornubian batholith which extends for 230 km from the Scilly Isles to Dartmoor. The batholith is exposed as a series of large bosses and minor cupolas with which the major Sn-Cu-W vein-style mineralisation of south-west England is associated. More gentle folding and faulting occurs to the north of the Variscan front (Figure 2) where Dinantian carbonates host major Pb-F-Ba mineralisation in the Northern and Southern Pennine orefields, North Wales and the Mendip Hills. Replacement hematite deposits formed near the contact of Dinantian carbonates and overlying Triassic sandstones in the western Lake District and South Wales.

3.7 Post-Palaeozoic to present

Post-Palaeozoic formations consist mainly of shallow-water, marine clastics and limestones devoid of significant mineralisation. The principal exceptions are Permo-Triassic basins containing red-beds with thick evaporites and extensive Jurassic sedimentary ironstones in the English Midlands. Red-bed Cu mineralisation occurs in the Triassic rocks of the Cheshire Basin. Economically significant ball clay deposits of Oligocene age occur associated with sands and lignite in a basin at Bovey Tracey in Devon. A large Tertiary basic to acid igneous province in north-west Scotland and Northern Ireland consists of extensive tholeiitic flood basalts and numerous major volcanic centres with associated high-level intrusions and dyke swarms. A thick sequence of varied Tertiary sediments, including lignite seams up to 140 m thick and averaging over 40 m in some areas, fills depressions in the basalt surface in Northern Ireland. Minor deposits of bauxite and laterite were formed during the contemporaneous weathering of basalt lava flows. The whole country, apart from the extreme south-west, was affected by Quaternary glaciation. This caused deep erosion in upland areas and has left widespread superficial deposits of sands and gravels which are exploited in most areas.

4 MINERAL EXPLORATION AND DISCOVERIES SINCE 1965

Mineral exploration of fluctuating intensity has continued in Britain over the last thirty years. A useful review of some of the earlier work is given in Howe (1982) and only a summary is presented here. A list of the more important discoveries and developments since the mid 1960s is given in Table 2 and their locations shown in Figure 3. In the mid 1960s new exploration techniques and metallogenic models

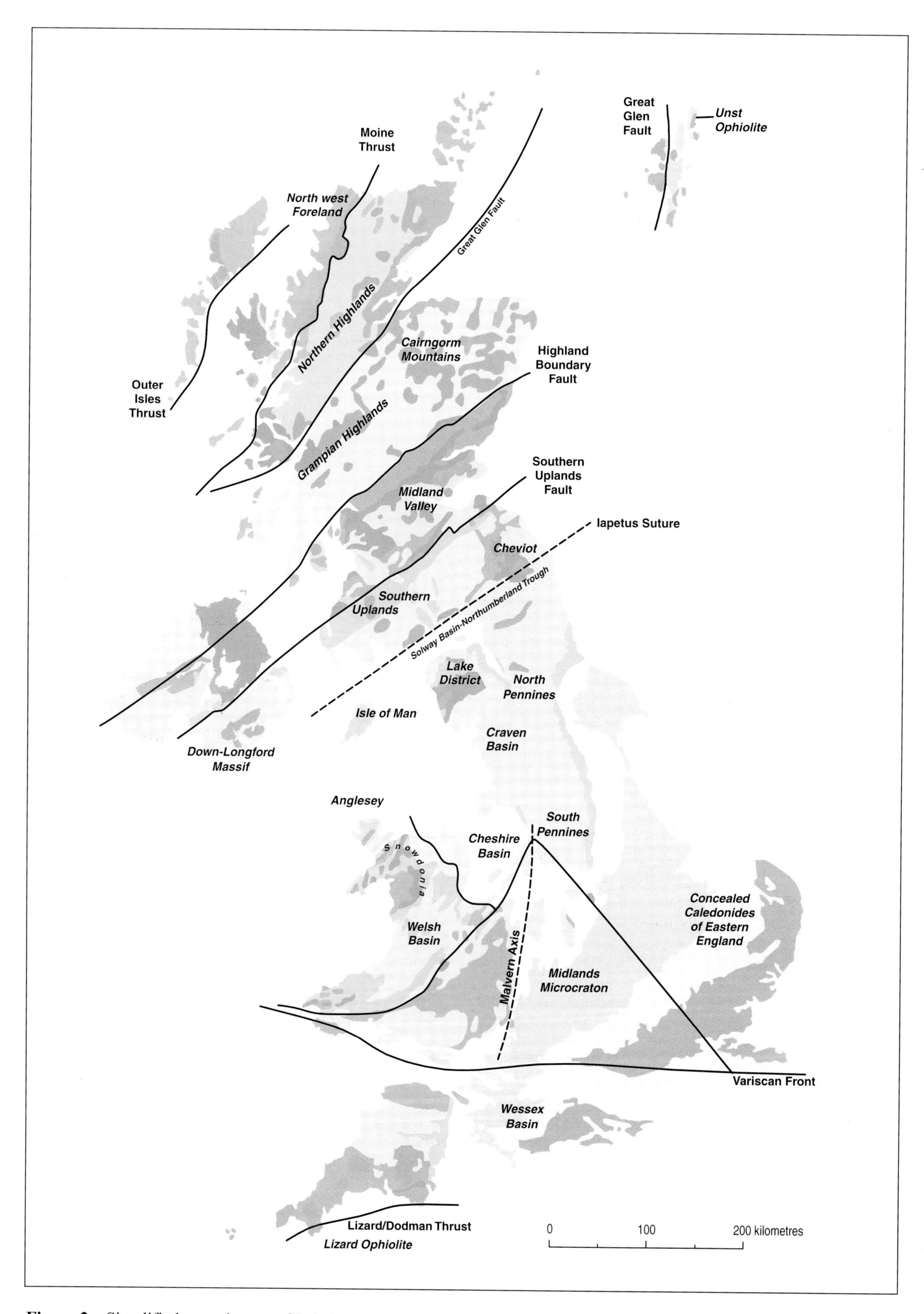

Figure 2 Simplified tectonic map of Britain.

stimulated a new look at Britain's mineral potential. A major programme was undertaken by Rio Tinto Finance and Exploration Ltd (RioFinex) between 1965 and 1973 resulting in the discovery of the Coed-y-Brenin porphyry copper deposit. A joint venture with Consolidated Gold Fields Ltd (later joined by Amax Exploration of UK Inc) in north-east Scotland identified Cu-Ni mineralisation associated with major Caledonian basic intrusions. Consolidated Gold Fields Ltd and other companies were also active in south-west England during this period. As a result, the Wheal Jane, Mount Wellington and Pendarves tin mines were brought on stream in the early 1970s. In 1974 the Boulby potash mine in North Yorkshire (Figure 3) began production, working Upper Permian Zechstein evaporites at a depth of 1100 m, and is currently producing around 1 000 000 t/y KCl.

The passing of the Mineral Exploration and Investment Grants Act 1972, under which the Department of Trade and Industry refunded up to 35% of exploration costs for approved programmes, gave further impetus to exploration activity. Several overseas companies were active at this time. They included Noranda-Kerr Ltd (especially in North Wales and northern Scotland), Phelps-Dodge Corporation NL (in North Wales and Scotland, especially for porphyry-style targets), Acmin Explorations (UK) Ltd (in the Pennine orefields), Swiss Aluminium Mining UK Ltd and the British Steel Corporation (in the Northern Pennine orefield).

Several underground and opencast fluorspar deposits (with associated baryte and galena) were developed in the Northern Pennine (Weardale mines) and the Southern Pennine (Sallet Hole and Long Rake mines) orefields. This exploration cycle culminated with the discovery of the major Aberfeldy baryte deposits in 1976 during the early stages of the Mineral Reconnaissance Programme (MRP) conducted by the BGS. The Foss deposit was subsequently developed by Dresser Industries Inc (now MI (GB) Ltd) while the larger, adjacent Duntanlich deposit, which contains at least 10 Mt of high-grade baryte, awaits planning consent for development.

Exploration in Northern Ireland was facilitated by the passage of the Mineral Development Act (Northern Ireland) 1969. A number of companies extended their interests from Carboniferous rocks in the Republic of Ireland, in which important base-metal discoveries had been made, to similar rocks in Northern Ireland, but without success. RioFinex examined the Ordovician volcanic rocks of County Tyrone for base-metals but found only minor intersections of low-grade copper mineralisation.

Exploration activity slowed during the late 1970s, though discoveries continued to be made. The Gairloch Cu-Zn-Au discovery by Consolidated Gold Fields in north-west Scotland was the first significant metalliferous ore deposit to be found in the Lewisian. Exploratory development at Hemerdon in south-west England by Amax Hemerdon Ltd demonstrated the presence of a major W-Sn deposit which could be worked profitably in favourable conditions. The continued investigation of the Parys Mountain deposit in Anglesey by Cominco (UK) Ltd provided fresh insights into the genesis and structure of this complex orebody. Several fluorite and baryte ventures flourished in the Pennine orefields. Increased tin and gold metal prices from 1980 onwards encouraged prospecting for these commodities and led to a second major cycle of exploration activity.

In south-west England a number of companies were active. South West Consolidated Minerals Ltd carried out an intensive programme of drilling in the Callington area and located a 'sheeted vein zone' of Sn-W mineralisation near the old Redmoor tin mine. Billiton Minerals UK Ltd were active both onshore for hardrock and alluvial tin prospects in the St Austell area, and offshore for alluvial deposits associated with buried river channels. Marine Mining (Cornwall) Ltd carried out trial working of placer tin deposits off the north Cornwall coast. Geevor Tin Mines Ltd developed decline access to the offshore sections of the mine and investigated several small prospects in Cornwall. Exploration of the Dalradian of Scotland for both baryte and base-metal mineralisation continued at a high level. The BGS, through the MRP, pursued its investigations along and across the strike of the Middle Dalradian rocks, which hosted the Aberfeldy stratabound baryte deposits. The MRP was also active in many other areas of Britain, looking for a number of styles of mineralisation. These included sedimentary stratabound targets in several areas, calc-alkaline porphyry deposits in Wales and southern Scotland, volcanogenic mineralisation in Wales and south-west England, carbonate-hosted base-metal mineralisation in the Pennines, ophiolite-hosted PGE mineralisation in Shetland and north-east Scotland and granite-related mineralisation in south-west England.

During the 1980s BP Minerals International Ltd investigated several areas for gold mineralisation associated with high-level acid intrusions in Scotland. These included Lagolochan, near Oban, and several localities in the Southern Uplands near Leadhills. BP also carried out drilling in the Craven Basin in northern England for carbonate-hosted Pb-Zn mineralisation. In the Northern Pennine Orefield, Minworth Ltd (through Weardale Holdings Ltd). acquired the fluorspar interests of British Steel and Swiss Aluminium. A Minworth subsidiary, Strontian Minerals Ltd, reopened the old Strontian lead mine in western Scotland to extract baryte from open pit and underground workings. Minworth also exploited baryte veins by open-pit operations in the Nethan Valley of central Scotland. In the Southern Pennine Orefield, Laporte Minerals Ltd continued to work the Sallet Hole fluorite mine and began exploitation of a lower-grade, replacement style of mineralisation in the Bradwell area. The company also developed the new Milldam mine at Great Hucklow. In North Wales, Clogau Gold Mines plc and a private venture at the Gwynfynydd Gold Mine carried out exploration and development work at old gold mine sites in the Dolgellau area. Exploration for, and production of, tin was curtailed by the onset of the 'tin crisis' in 1985. Two mines, Wheal Jane and South Crofty, were kept open with assistance from the Department of Trade and Industry, but the continuing depressed tin price caused the closure of Wheal Jane in 1991 and South Crofty in 1998.

Exploration in the mid to late 1980s focused on gold in the Dalradian, following the discovery by Ennex International plc of the Cononish and Curraghinalt gold deposits in Scotland and Northern Ireland respectively. New discoveries included vein-style mineralisation at Calliachar/Urlar Burn, near Aberfeldy by Colby Gold plc and the Cavanacaw deposit west of Omagh in Northern Ireland, which was found by RioFinex. The MRP continued to find significant discoveries of baryte and base-metal mineralisation in the Dalradian, gold mineralisation in Devon associated with Permo-Triassic red-beds, evidence of volcanogenic mineralisation in south-west England and Wales, and PGE in Shetland and north-east Scotland. In 1988–90 Anglesey Mining plc sank the 300 m deep Morris shaft and carried out detailed underground drilling and 1 km of underground development to investigate the western end of the Parys Mountain Cu-Zn-Pb-Ag deposit. Reserves were delineated and a pilot plant set up to develop satisfactory milling and treatment processes. Falling metal prices forced the suspension of the project in 1991 but further development work has recently taken place.

Figure 3 Metalliferous mineral prospects discovered or developed since 1965.

In the 1990s gold exploration continued in Scotland with the underground exploration of the Cononish deposit (Figure 4) and the drilling of the Rhynie epithermal mineralisation. The BGS MRP discovery of widespread alluvial gold in the South Hams area of Devon led to the recognition of a new style of gold mineralisation associated with Permian red-bed sequences, especially where alkali basaltic volcanism is present. A number of Permo-Triassic basins were found to contain evidence of red-bed sourced gold mineralisation and several companies are investigating the potential of this style, most notably in the Crediton Trough, near Exeter, where a number of boreholes have been drilled. Anglesey Mining restarted exploration at the Parys Mountain deposit in 1997 and a number of additional boreholes were drilled to test a new mineralisation model. Anglesey Mining is also investigating turbidite-hosted gold mineralisation adjacent to the old Ogofau gold mine in mid-Wales. Renewed interest has also been shown in copper-nickel and PGE prospects in north-east Scotland and Unst, stimulated in part by the Voisey's Bay discovery and demand for PGEs.

Following publication of a MRP report on the potential for diamonds in Britain, companies started exploration for indicator minerals in northern Britain and Ireland, as no kimberlites are known to outcrop. No diamonds have yet been found, but a number of indicator minerals have been located in the north of Ireland. The MRP continued to publish reports detailing mineral discoveries in a number of areas, mainly for gold, until 1998 when it was subsumed into a new Minerals Programme which is concentrating on regional appraisals and the dissemination of information.

5 FUTURE PROSPECTS FOR MINERAL EXPLORATION

The work outlined in the previous section together with other studies by BGS and University-based researchers indicates that Britain has potential for the following styles of mineralisation. These styles are recognised elsewhere in the world and their geological settings and exploration criteria are well established:

- Sedimentary exhalative deposits of baryte, base-metals and gold.
- Mesothermal gold deposits, predominantly of slate-belt (turbidite-hosted) type.
- Unconformity-related (redox) gold deposits.
- Volcanogenic deposits of base-metals, silver and gold, including volcanic massive sulphide deposits.
- Carbonate-hosted deposits of Irish and Pennine sub-types.
- Mafic-intrusion-related nickel-copper deposits.
- Epithermal gold deposits associated with Devonian volcanism.
- Gemstones in deep seated minor intrusions.
- Granite-related polymetallic vein and stockwork mineralisation, principally for tin and tungsten.
- Calc-alkaline porphyry-style mineralisation.
- Placers.
- Platinum Group Elements (PGE) associated with alkaline intrusions.
- Vein-style basin-related base-metal deposits.

5.1 Sedimentary stratabound mineralisation

5.1.1 Sedimentary exhalative (SEDEX) mineralisation

The main zone of interest is the Scottish Middle Dalradian (Argyll Group) sequence from Portsoy in the north-east to Islay in the south-west with the Aberfeldy Ba-Pb-Zn deposits (Figure 3) in the centre (Hall, 1993). A detailed helicopter EM survey was flown over extensive areas of the Middle Dalradian in the early 1980s (MEG 253), but with only limited ground follow-up.

There are a number of highly prospective areas for Ba-Pb-Zn mineralisation recorded in the MRP reports listed below. The mineralisation is of SEDEX type (Large, 1983) with stratabound sulphates, silicates and sulphides occurring in small, third order basins within the Middle Dalradian (Figure 4) and especially in the Ben Eagach Schist Formation (Coats et al., 1980). Lenses of Pb-Zn sulphides with up to 10% Pb+Zn have been found at several localities, and areas of barium-enriched sediments also occur (Coats et al., 1984a).

Two horizons of sulphidic quartzite with Zn-Cu-Pb mineralisation occur in the Tyndrum area of western Scotland (Figure 28) within the Argyll Group above the Ben Eagach Schist Formation which hosts the Aberfeldy mineralisation (Smith et al., 1984). The upper Ben Challum horizon is up to 20 m thick with a strike length of several kilometres. Channel sampling has proved up to 3% Zn and 0.1% Pb over one metre. The lower Auchtertyre horizon is about 80 m thick over a strike length of 8 km and contains up to 1.6% Zn and 0.1% Cu over 3 m (MRP 93; Fortey and Smith, 1986). Mineralisation at the western end of the Ben Challum horizon has been attributed to hydrothermal systems associated with the development of basic intrusive rocks at shallow depth forming a possible Besshi-style deposit (Scott et al., 1988). Mineral reconnaissance by the BGS north-east of Aberfeldy located a number of geochemical and geophysical anomalies in the Ben Eagach Schist Formation around Glenshee (Pease et al., 1986) where Zn-Pb-Ba mineralisation has been found (MRP 88). Exploration at Loch Kander (near Braemar) found bedded baryte up to 5 m thick, with significant base-metal values, over a 700 m strike length in the same formation (MRP 104).

The Argyll Group outcrop from Braemar north to the coast at Portsoy has been less explored than its continuation to the south-west, because of the complexity of the structure, poor exposure along strike and lithological variation. However, MRP exploration has found stratabound Pb-Zn enrichment in quartzites and pelites in the Wellheads area of the Upper Deveron Valley (MRP 145). A large EM and magnetic anomaly was reported by the EVL in the Creagan Riabhach area north of Ballater over calc-silicates and semipelites associated with metabasite intrusions. Follow-up surveys (MRP 87 and 145) have related this anomaly to a sulphidic (pyrite/pyrrhotite) horizon, resembling the Pyritic Zone in the Ben Lawers Schist Formation. This persistent horizon has now been traced for 250 km from Glenshee to Loch Fyne (Smith et al., 1984), but it is largely devoid of base-metals except in the south-west, where drilling in the Meall Mhor area of Knapdale intersected pyritic quartzite with up to 1% Cu over 2.7 m (MRP 15). Mining companies, including Consolidated Gold Fields Ltd (MEG 4) and Noranda-Kerr (MEG 74 and 115) have carried out exploration in the Knapdale area and more recent MRP work has revealed the presence of gold mineralisation (Section 5.2.1).

Interest in the Dalradian in Northern Ireland, already under active exploration for mesothermal vein-style gold deposits, has been increased by the unexpected discovery of hydrothermally altered Dalradian metasediments, with associated base-metal sulphides, in boreholes drilled by Meekatharra Minerals through Tertiary basalt cover south of Ballymoney. The basalt was previously estimated to be 2 km thick but the boreholes intersected mineralised Dalradian rocks at a depth of <100 m. Legg et al. (1985) have published an index and map of mineral localities in the Dalradian of Northern Ireland and part of the Republic of Ireland.

Table 2 Principal discoveries and developments in Britain since 1965.

Location	Commodities	Type	Activity
Arthrath, northeast Scotland	Copper-nickel	Basic to ultrabasic intrusions of Caledonian age in pyritic and graphitic Dalradian metasediments. Copper and nickel sulphides developed at or near the contacts	Discovered by RioFinex in the late 1960s. Complex structure makes drill hole correlation and ore reserve calculation difficult. Resources of 17 Mt at 0.21% Ni and 0.14% Cu. Exploration data are on Open File in MEG 21.
Calliachar Burn and **Urlar Burn**, Aberfeldy, Scotland	Gold	Multiple small quartz-sulphide veins in Dalradian metasediments	Discovered by Colby Gold plc in 1988. Grades of individual veins up to 8.81 g/t over a strike length of 87.5 m. Under investigation as a source of small amounts of gold for jewellery.
Cavanacaw, Omagh, Northern Ireland	Gold	Mesothermal 20 m wide quartz vein zone in Dalradian metasediments in the Sperrin Mountains, near Omagh, Northern Ireland	Discovered by RioFinex in 1987. Total resources of 2 Mt at 6.9 g/t Au. Full planning consent has been granted and the current owner (Omagh Minerals Ltd, a wholly owned subsidiary of European Gold Resources Inc). has announced plans to commence development in 2000.
Coed-y-Brenin, North Wales	Copper (gold)	Porphyry-style deposit in Cambrian intrusives and sediments	Discovered by RioFinex in 1968. Resources have been estimated at about 200 Mt at 0.3% Cu with minor gold. Currently inactive and within the Snowdonia National Park. Some exploration data are on Open File in MEG 5.
Cononish, western Scotland	Gold	Mesothermal quartz-sulphide vein structure in Dalradian metasediments	Discovered by Ennex in 1984. Veins extend over one kilometre strike length. A mineable resource of 514 000 t at 9.42 g/t Au (cut and diluted) and 53 g/t Ag has been announced, with some high-grade intersections. Planning consent for a mine has been granted and the current owners (Caledonia Mining Corporation) are awaiting favourable economic conditions to proceed to full production.
Credition Trough, south-west England	Gold	Unconformity-related mineralisation in a narrow graben of Permo-Triassic red-bed sediments with alkali-basalt lavas	Gold in surficial deposits discovered in 1991 by the MRP. Now under investigation by Crediton Minerals who have drilled a number of shallow boreholes. MRP 133 and 134 provide details of the discovery.
Curraghinalt Omagh, Northern Ireland	Gold	Mesothermal quartz-sulphide veins in Dalradian metasediments	Discovered by Ennex in 1983. Geological resource is currently 470 000 t at 17.3 g/t Au. Awaiting favourable economic conditions to proceed to full production. Now owned by Nickelodeon Minerals Inc.
Foss and **Duntanlich**, Aberfeldy, Scotland	Baryte	SEDEX (sedimentary exhalative) stratiform syngenetic, high-grade baryte deposit in Dalradian metasediments with associated lead and zinc sulphides	Discovered by the BGS in 1976. MRP data relating to the discovery of the deposit are published in MRP Reports 26 and 40. Now owned by MI (GB) Ltd. The Foss deposit is currently in production for baryte at a rate of about 50 000 t p.a.). The larger Duntanlich deposit is awaiting planning permission. Total resources of serveral million tonnes of baryte.
Gairloch, north-west Scotland	Copper-zinc-gold	Stratiform, volcanogenic Besshi-style massive sulphide. Low-grade, Cu-Zn deposit with associated gold values in Lewisian sediments and volcanics	Discovered by Consolidated Gold Fields in 1978. Estimated resource of about 500 000 t at 1% Cu, 0.5% Zn and 1 g/t Au. Currently inactive. Exploration data are on Open File in MEG 173.
Hemerdon, south-west England	Tungsten-tin	Low-grade, W-Sn stockwork deposit in Variscan granite	Published reserves of 42 Mt at 0.2% W+Sn. First major production 1915. Modern evaluation from 1976 by Amax Hemerdon. Currently inactive, but with planning permission for development as a 2.2 M t/y open pit mine. Exploration data are on Open File in MEG 188.

Lagalochan (Kilmelford) western Scotland	Copper-gold	Porphyry-style desposit in calc-alkaline grandiorite intrusions of Caledonian age with zones of hydrothermal alteration, brecciation and disseminated sulphides	Kilmelford was discovered by mining companies in 1971 and drilled by BGS in 1976. Very low-grade (less then 0.1% Cu) copper mineralisaion reported in MRP report 9. An adjacent area, Lagalochan, was investigated as a gold prospect by BP Minerals in the 1980s. Intensive drilling located numerous strongly altered and brecciated zones, some with associated gold. Currently inactive. Exploration data are on Open File in MEG 258.
Littlemill nr Huntly north-east Scotland	Copper-nickel	Basic to ultrabasic intrusions of Caledonian age intruding pyritic and graphitic Dalradian metasediments. Low-grade Cu - Ni sulphides developed at or near the contacts	Discovered by Consolidated Gold Fields in the late 1960s. Complex structure makes drill hole correlation and ore reserve calculation difficult. Resources of 2 Mt at 0.52% Ni and 0.27% Cu. Exploration data are on Open File in MEG 8 and 110.
Parys Mountain North Wales	Copper-zinc-lead-silver	Volcanogenic massive sulphide deposit in acid and basic volcanics and sediments at the Ordovician — Silurian boundary	Discovered in 1768 and worked in a series of open pits and underground mines to the 1800s. Re-interpreted in the 1970s as a volcanogenic massive disseminated sulphide deposit. Currently under development by Anglesey Mining plc. Drill-indicated reserves of 4.8 Mt at 1.5% Cu, 3.0% Pb, 6.0% Zn, 57 g/t Ag and 0.4 g/t Au plus larger resources of lower-grade copper mineralisation.
Redmoor, south-west England	Tin-tungsten	Sheeted vein deposit in Devonian metasediments associated with Variscan granite	Discovered by South West Consolidated Minerals in 1981. Multi-million tonne 'resource'. Exploration data are on Open File in MEG 165.
South Crofty, south-west England	Tin	Multi-stage swarms of quartz-cassiterite fissure veins associated with Variscan granite near the contact with Devonian metasediments and basic volcanics	Worked for over 100 years before closure in 1998. The final year's production was 192 544 t of ore at a head grade of 1.41% Sn producing 2395.8 t of tin in concentrates.
Unst, Shetland, Scotland	PGE	Hydrothermal and magmatic PGE in ophiolitic ultrabasic complex	Associated with an area of old chromite workings. Investigated by BGS from 1983 to 86; data are available in MRP 73. Investigated by Leicester Diamond Mines Ltd. in 1999.
Vidlin, Shetland, Scotland	Copper-zinc	Stratiform, volcanogenic massive sulphide. Small, low-grade Cu-Zn deposit within a zone up to 4 km long and 10 m thick in Dalradian metasediments	Investigated from 1973. BGS MRP exploration data on the discovey of the deposit are available in MRP 4 and company investigations by Messina (Transvaal) are on Open File in MEG 150.
Wheal Jane, south-west England	Tin-zinc-copper-silver	Complex Sn-Zn-Cu-Ag-As sulphide deposit at elvan (quartz porpyry) dyke contact with Devonian metasediments	Developed as a modern mine in 1971 on the site of old workings producing tin with co-product zinc and by-product copper and silver. Closed in 1991.

In south-west Scotland, stratiform Au-As-Sb mineralisation occurs at the Glendinning mine in Lower Palaeozoic greywackes (MRP 59). Additional studies on this deposit are reported by Duller et al. (1997).

Stratabound Pb-Zn-Ba mineralisation has been found in Devonian and Carboniferous sedimentary and volcanic rocks at several sites in south and west Devon. A 1–2 m thick black shale horizon with about 2% Pb-Zn (but 10% Pb over 4 m of drill core in a fold nose) was found by RioFinex at Egloskerry (Figure 3) near Launceston (MEG 223). Minor Pb-Zn-Ag mineralisation was discovered by South West Consolidated Minerals Ltd at the Devonian-Carboniferous boundary in the Callington area north-west of Plymouth (MEG 212), and the MRP reported disseminated and veinlet base-metal mineralisation in a condensed Carboniferous succession at Higher Coombe in the Teign valley. Highest values in drill holes at the latter site were 2% Zn over 3 m with weaker Cu and Pb (MRP 123). Some of the mineralisation at these sites is associated with strong alteration of the enclosing sediments and tuffs and all could be of volcanogenic origin.

MRP investigations south of the Dartmoor Granite in the Middle Devonian volcano-sedimentary succession proved 7 m of massive pyrite in drill core with associated high-grade baryte float (MRP 79). Additional exploration showed further evidence of metal-enriched (Zn-Pb) sedimentary rocks and indicated that extensive hydrothermal systems were associated with the alkali basalt volcanism in the area (MRP 129). The whole Devonian-Lower Carboniferous volcano–sedimentary belt in south-west England has considerable exploration potential for strati-

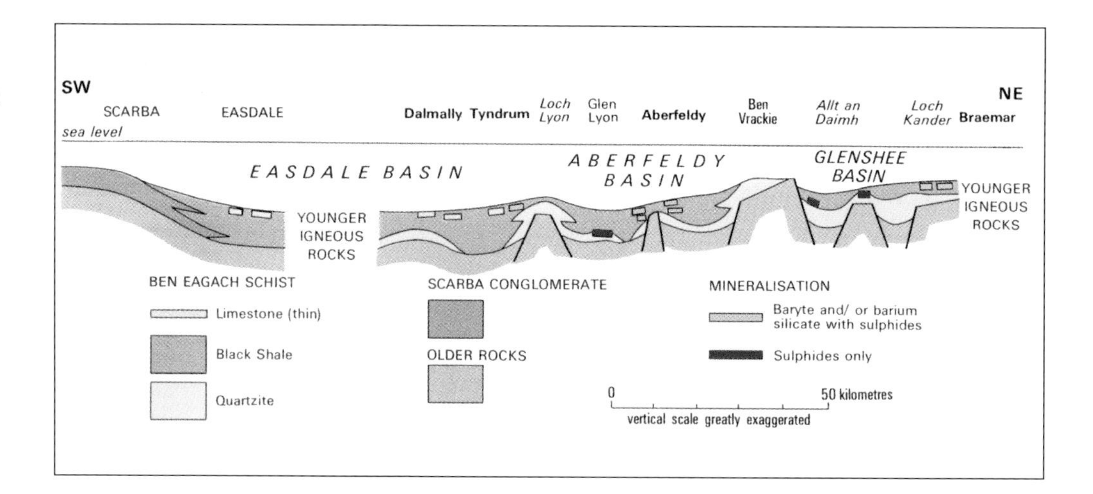

Figure 4 Schematic cross-section of the Dalradian basin showing the location of known mineralisation.

form SEDEX mineralisation of the baryte-base-metal type similar to the Meggan or Rammelsberg deposits of equivalent age in the Federal Republic of Germany. Exploration is hampered by poor exposure and the rocks have been strongly tectonised and deeply weathered. Geochemical soil traverses and deep overburden drilling have proved useful exploration techniques (MRP 129).

A more speculative exploration area for stratabound mineralisation is the south-eastern margin of the back-arc Lower Palaeozoic Welsh Basin where it abuts the continental margin of the Midlands Microcraton. Badham (1981) has suggested that the basinal Lower Silurian shales to the west of the carbonate shelf facies could have formed a suitable depositional environment for SEDEX mineralisation, and Haggerty et al. (1996) consider that the source of the Ba-Pb-Zn vein mineralisation hosted by Lower Ordovician shales in the Shropshire orefield could have been basinal brines from the Welsh Basin. Bevins (1985) has described hydrothermal alteration in Ordovician lavas in the Builth Wells area (Figure 28) and exploration by the BGS, including shallow drilling, located stratabound lead mineralisation in the volcanic succession (MRP 92).

In North Wales, an extensive though generally steeply dipping 30–40 cm thick manganese ore bed of Cambrian age has been worked from many small opencast mines in the Harlech Dome (Bennett, 1987). The ore consists of a fine intergrowth of rhodochrosite and spessartine with disseminated hematite and pyrolusite, and there is little doubt that considerable resources remain at depth. Younger deposits, of probable volcanic-exhalative origin, have been exploited from Ordovician mudstones and dolerites at Rhiw in the Lleyn Peninsula. Intense Caledonian shearing and faulting have resulted in a series of lenticular orebodies up to 9 m thick. The MRP investigated this area using magnetic surveys and drilling, but only minor new mineralisation was found (MRP 102). Further, smaller occurrences were worked in Ordovician tuffs at Arenig.

5.1.2 Sediment-hosted copper deposits in post-orogenic basins

The Triassic Sherwood Sandstone Group of central England is frequently cemented by baryte. The baryte is often associated with minor red-bed copper mineralisation, as at Alderley Edge in Cheshire (Figure 23), adjacent to faults as shown in Figure 5 (Warrington, 1966; Carlon, 1979). These have provided pathways for fluids generated during diagenesis of the fault-bounded Triassic basins (Holmes et al., 1983; Naylor et al., 1989). Some extensive outcrops of sub-horizontal, baryte-cemented sandstone occur and could be easily worked but the grade is very variable and usually low. MEG 208 contains data on baryte exploration, including shallow drilling, in Cheshire. A recently completed multidisciplinary study of the Cheshire Basin (Plant et al., 2000) developed a metallogenic model involving remobilisation of metals by breakdown of primary minerals, scavenging of metals from mudstones in the upper part of the basin fill associated with the flow of density driven brines, influx of reducing fluids along basin-margin faults causing ore precipitation, and alteration of ores by meteoric water following basin inversion (Figure 5). Carbon isotope signatures of diagenetic carbonates appear to distinguish areas of potential copper mineralisation from zones of dominantly iron sulphides.

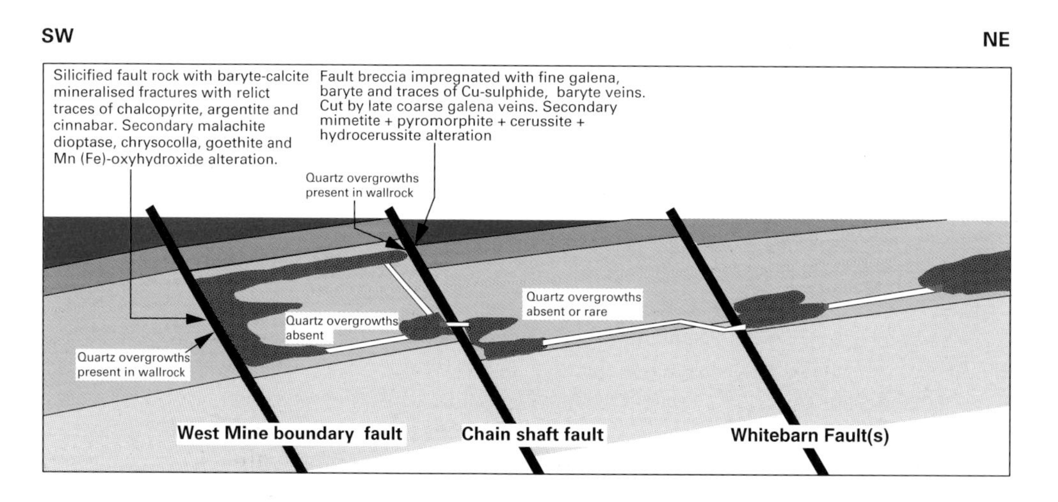

Figure 5 Alderley Edge mine, Cheshire. Diagrammatic illustration of the relationship between faults and mineralised sandstone bodies in West Mine, Alderley Edge, Cheshire (Plant et al., 2000). Extent of copper mineralisation in dark grey.

5.1.3 *Sediment-hosted uranium deposits in post-orogenic basins*

In the early 1970s the Devonian Orcadian basin in northern Scotland was found to contain local generally low-grade concentrations of uranium (Michie and Cooper, 1979; Michie and Gallagher, 1979). The uranium is mostly found close to the base of the Middle Devonian succession, notably associated with arkosic breccias marginal to the Helmsdale granite on the east coast of Caithness at Ousdale, in a fault breccia at Mill of Cairston near Stromness on Orkney (McCready and Parnell, 1998), and in small amounts at many sites associated with phosphate and hydrocarbons in the carbonate-rich shale, siltstone and fine sandstone cyclic succession. The Ousdale and Mill of Cairston occurrences were drilled by the BGS and mining companies. Maximum values of 0.1% U and 5.5% Pb were found at the Mill of Cairston (MEG 125) and 850 g/t U within a 15 m intersection at Ousdale (MEG 82). The rocks of the basin also contain lenses of stratiform baryte and manganese, as at Balfreish, south-east of Inverness (Nicholson, 1990).

Uranium and vanadium concentrations occur in nodules within the red-bed dominated Permian succession near Budleigh Salterton in South Devon. They have been the subject of several investigations, including drilling (MRP 89).

Relevant MRP and MEG reports containing information relevant to stratabound mineralisation include:

- Dalradian: MRP 8, 13, 15, 23, 26, 37, 40, 50, 87, 88, 93, 104 and 145; MEG 2, 3, 4, 74, 82, 104, 115, 123, 125, 126, 243 and 248.
- South-west Scotland: MRP 59, 69; MEG 2, 3 and 230.
- Wales: MRP 92.
- Central England: MRP 52; MEG 208.
- South-west England: MRP 79, 89, 123 and 129; MEG 73, 100, 140, 212, 223 and 235.

5.2 Mesothermal lode gold mineralisation

5.2.1 *Scotland*

Since the early 1980's several important prospects have been found in the Dalradian rocks of the central and south-west Highlands. The most important known deposit of this type is situated at Cononish (Figure 3) in the historic Tyndrum lead-mining district, close to a major north-east-trending structure, the Tyndrum Fault.

Figure 6 Adit development at the Cononish gold prospect, south-west Highlands, Scotland.

The deposit comprises a steeply dipping quartz vein, up to 6 m wide, cutting Lower Dalradian psammites (Earls et al., 1992). It can be traced along strike for over 1 km and over a vertical distance greater than 250 m. Underground exploration has defined a resource of 483 000 t at 15.9 g/t Au and the company has planning permission for development of an underground mine (Figure 6).

There are many recorded occurrences of gold in bedrock and alluvium in rocks of similar age between Aberfeldy and Comrie. Follow-up of anomalous values of gold in heavy mineral pan concentrates reported by the BGS led to the discovery in the late 1980s of gold-bearing vein structures at Calliachar Burn, 4 km south-west of Aberfeldy. The quartz-carbonate veins, accompanied by extensive wallrock alteration, are preferentially developed in quartzites and well-jointed amphibolites, with gold grades in one structure averaging 8–9 g/t over a strike length of 87 m (Ixer et al., 1997). The overlying gossan was reported to contain up to 400 g/t Au and 230 g/t Ag. Gold is also reported from four narrow veins in the nearby Urlar Burn. Other occurrences in the south Loch Tay area are located in a stockwork of galena-bearing quartz veins at Corrie Buie and in a shear zone cutting the southern margin of the Comrie diorite complex (MEG 226). In the Pitlochry to Glen Clova district of the central Highlands follow up of geochemical anomalies and panned gold observations have identified several targets. Gold concentrations up to 6.8 g/t were reported in a fault zone (MRP 126).

Recent MRP work has indicated the potential for economic lode gold mineralisation in the extreme south-west of the Dalradian outcrop in Scotland (MRP 143). Gold contents in the ppm range have been reported from base-metal-bearing quartz-carbonate veins in metasediments of the Knapdale area and nearby at Meall Mor gold enrichment occurs in association with stratabound copper mineralisation.

The BGS has carried out a survey for gold over parts of the Shetland Islands in the north of Scotland and located a number of potential target areas within the Middle to Upper Dalradian succession (Buchanan and Dunton, 1996). In the Southern Uplands of Scotland mesothermal gold-bearing veins are associated with late-Caledonian granitic intrusions and major strike-slip faults. The MRP found quartz veins containing up to 1.5 g/t Au over 4.5 m in drill core at Glenhead Burn (MRP 46), and BP Minerals carried out an intensive investigation for gold associated with granitoid intrusions at Hare Hill (Boast et al., 1990), Moorbrock Hill, Stobshiel and Glenhead Burn (Naden and Caulfield, 1989). Hydrothermal alteration and mineralisation continued for up to 30 Ma after the emplacement of the intrusions at about 400 Ma. Gold mineralisation is associated with earlier fluids; antimony and base-metal sulphides with later fluids. Diagenetic pyrite in the Lower Palaeozoic rocks of the Southern Uplands provided a sulphur reservoir (Lowry et al., 1997). Additional information is given in MEG 257, 259, 260 and 261.

In the Leadhills-Wanlockhead area of southern Scotland, where Pb-Zn vein deposits and alluvial gold have been worked historically, exploration by companies identified several new occurrences of gold in bedrock. In the Duns area the MRP identified gold in stream sediments and bedrock associated with major structural features and small intrusions (MRP 138). Minor As-Sb-Au mineralisation of mesothermal origin has been documented at a former antimony mine in Silurian greywackes at Glendinning, near Langholm (Duller et al., 1997).

A major review and multidataset analysis of Caledonian and Variscan gold mineralisation in selected areas of western Europe (the Midas Project) provides a large amount of useful information and metallogenic models to assist in gold

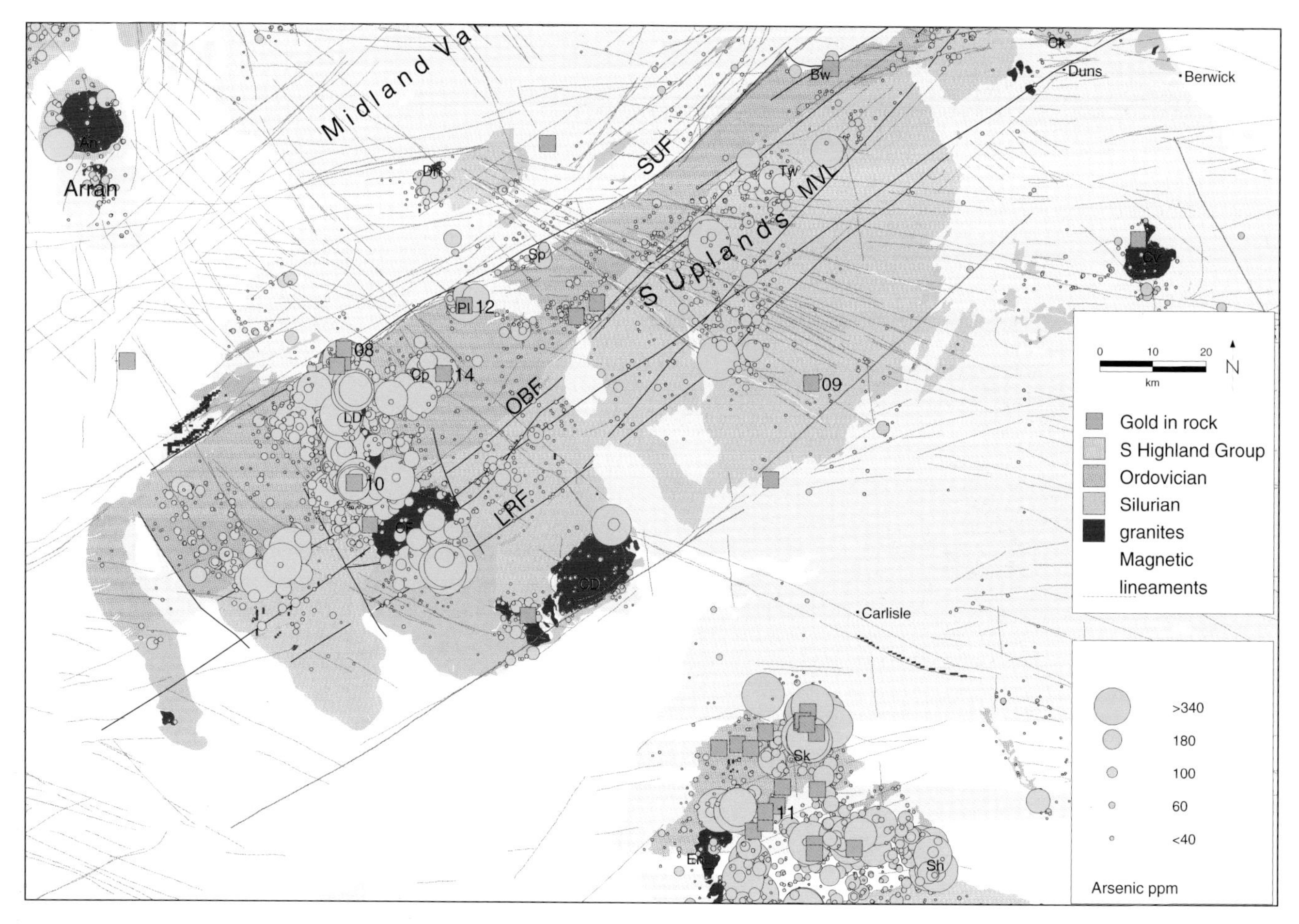

Figure 7 GIS analysis for gold exploration in the Southern Uplands of Scotland. Red areas mark the outcrop of granite intrusions. From Plant et al., (1998).

exploration in the region (Plant et al., 1998). In Britain, the Lagolochan copper-gold porphyry system and the mesothermal gold veins in the Southern Uplands of Scotland were investigated in some detail. Figure 7 shows the application of mesothermal gold mineralisation model criteria to integrated digital geological, geochemical, geophysical, structural and mineral occurrence data in the Southern Uplands. This approach was also used in a recent MRP multi-dataset analysis of the Southern Uplands (MRP 141) to identify several favourable areas for gold mineralisation which had not been investigated previously. Test field sampling in a few of the areas revealed new indications of gold mineralisation suggesting that the methodology was appropriate and that further exploration was merited.

Further studies of gold exploration methods in Britain, including the use of artificial intelligence, concentrated on models demonstrating the prospectivity for mesothermal lode gold in the Dalradian terrane of central and western Scotland (Gunn et al., 1997). Figure 8 shows a prospectivity prediction around the Cononish area, based on a particular set of criteria (Model G2), which assigns weightings to various elements of the model, including gold occurrences, proximity to lineament intersections, position of geophysical features etc. The highest prospectivity (0.8 to 1.0) is shown in red, the lowest in blue. This work has now been extended to investigate the potential for epithermal gold mineralisation in the Devonian volcanic rocks of Scotland. Additional information on the use of information technology in mineral exploration is given in Hatton et al. (1998).

5.2.2 Northern Ireland

Interest in the potential of the Dalradian of Britain for gold mineralisation was sparked off by the discovery of gold in 1983 by Ennex International plc at Curraghinalt (Figure 3) in the Sperrin Mountains north-east of Omagh in County Tyrone (Clifford et al., 1990). A number of east-west striking quartz vein structures were sampled and found to be auriferous. The veins were tested initially by trenching and then by over 60 diamond drill holes which produced a resource estimate of 900 000 t at 9.60 g/t Au (Clifford et al., 1992). An adit was driven to test a 400 m long section of the main structure. A further 10 000 m drilling programme was undertaken in 1997–98 to upgrade the resource classification from an inferred/indicated resource to an indicated/measured resource prior to the commencement of a feasibility study. The Curraghinalt deposit, and two associated exploration licences, were acquired by Nickleodeon Minerals Inc. in 1999 and resources are now stated to be 470 000 t at 17.3 g/t Au.

Following the Curraghinalt discovery a number of other companies took out licences in the area. Riofinex North selected the Lack Inlier, to the south-west of Omagh, and quickly found high-grade auriferous quartz vein float (Cliff and Wolfenden, 1992). Further exploration using deep overburden and diamond drilling under thick drift cover located the north-south-striking Kearney Structure which hosts the Cavanacaw (formerly called the Lack) deposit (Figure 3). The Kearney Structure, which is around 5 m wide, contains complex brecciated, sub-vertical quartz veining with pyrite and galena. The vein was exposed by trenching and over 2000 channel samples taken to give a mean grade of 7.6 g/t Au, 19.9 g/t Ag and 0.9% Pb (Cliff and Wolfenden, 1992). The

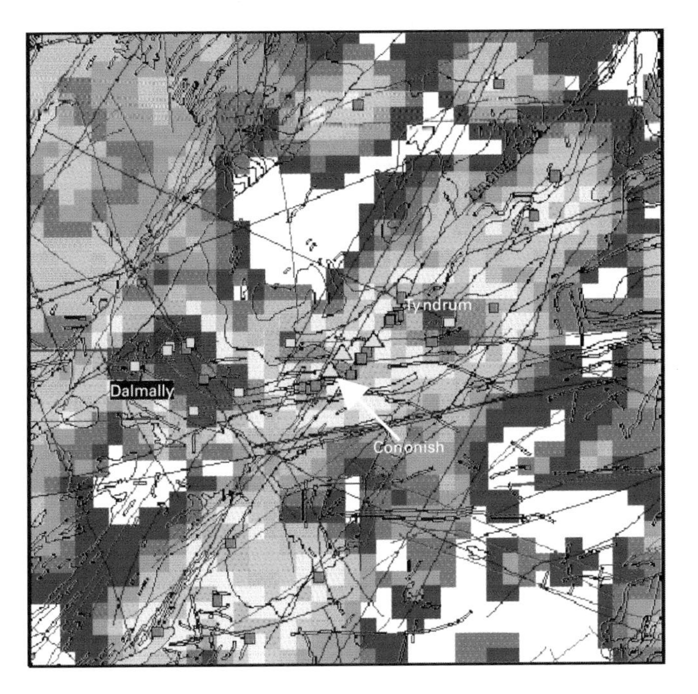

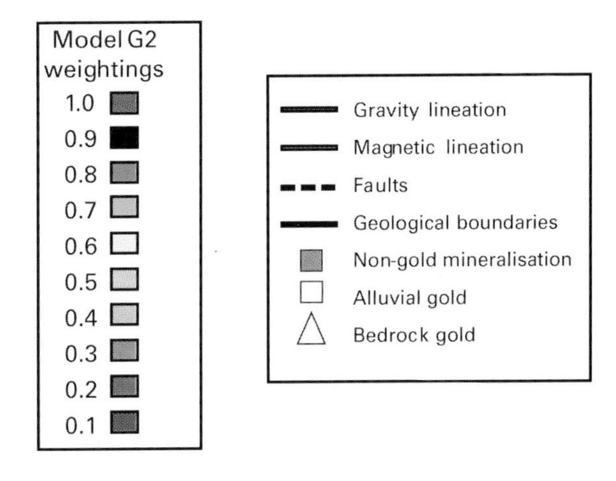

Figure 8 Gold prospectivity map for the Cononish area, western Scotland.

prospect is now owned by Omagh Minerals, a wholly owned subsidiary of European Gold Resources Inc, who have obtained full planning permission for development of an open pit mine following an extensive public enquiry. Total resources are now estimated at 2 Mt at 6.9 g/t Au. Initial production is planned to start in 2000.

5.2.3 Lake District

Mineral exploration by the MRP in the Black Combe area of the southern Lake District (Figure 24) found evidence of vein-style, turbidite-hosted gold-bearing polymetallic mineralisation (MRP 128). The mineralisation is hosted by Ordovician siltstones and mudstones of the Skiddaw Group within the Westmorland Monocline, a major Caledonian structure at the south-east margin of the Lake District batholith interpreted as an Acadian mountain front. The prospective area may extend some distance to the north-east within this regional structure.

5.2.4 Wales

The Harlech Dome in southern Snowdonia is rich in metalliferous minerals, and the mines of the Dolgellau 'gold belt' (Appendix 1), on the south and east sides of the dome, have dominated gold production in the British Isles since reliable records have been kept. In this area gold occurs in fault-controlled polymetallic quartz-sulphide veins cutting fine to coarse grained clastic sedimentary rocks of Cambro-Ordovician age deposited in the Welsh Basin. Usually the veins only contain appreciable gold where they intersect graphitic horizons within the clastic succession, notably the black mudstones of the Clogau Formation. (Shepherd and Bottrell, 1993).

About 75 km to the south lies the Ogofau deposit (Figure 24), where gold occurs in pyritic shales, quartz veins and reefs within a tightly folded and sheared clastic sedimentary succession, in the Lower Palaeozoic Welsh Basin, close to the Ordovician-Silurian boundary. In both areas the mineralisation has features in common and is typical of turbidite-hosted mesothermal deposits (Annels and Roberts, 1989). The precise reason for economic gold deposits apparently only occurring at these locations is still not clear, and there are some indications from regional studies that undiscovered deposits of the same style may exist elsewhere in the Welsh Basin where suitable structural and lithological conditions are fulfilled (MRP Data Release 13).

Traces of gold of probable mesothermal style have been recorded from several other localities in Wales; notably in northern Anglesey associated with quartz vein structures developed along fault and thrust boundaries in late Precambrian and Lower Palaeozoic rocks.

5.2.5 South-west England

In south-west England, exploration in Middle Devonian volcanic and sedimentary rocks north of Wadebridge in north Cornwall has located up to 1 g/t Au associated with quartz-arsenopyrite-stibnite veins (Clayton et al., 1990; MRP 103). Further gold occurrences in the region suggest that the potential of the Variscan rocks to host gold deposits merits further attention. Camm (1995) suggests that over one tonne of gold may have been recovered during the working of the alluvial tin deposits in the region.

Relevant MRP and MEG reports containing information relevant to mesothermal gold mineralisation include:

- Dalradian MRP 50, 126, 143; MEG 226, 233, 248.
- Southern Scotland: MRP 138, 141; MEG 257, 259, 260, 261.
- Wales and south-west England: MRP 99, 103, 112 and Data Releases 13 and 14.

5.3 Unconformity-related (Redox) gold deposits

The discovery of a close association between alluvial palladium-rich gold and Permian red-bed sediments and volcanics in south Devon by Leake et al. (1991) during MRP work led to reconnaissance surveys across most of the Permo-Triassic outcrop in Britain. As a result, a mineralisa-

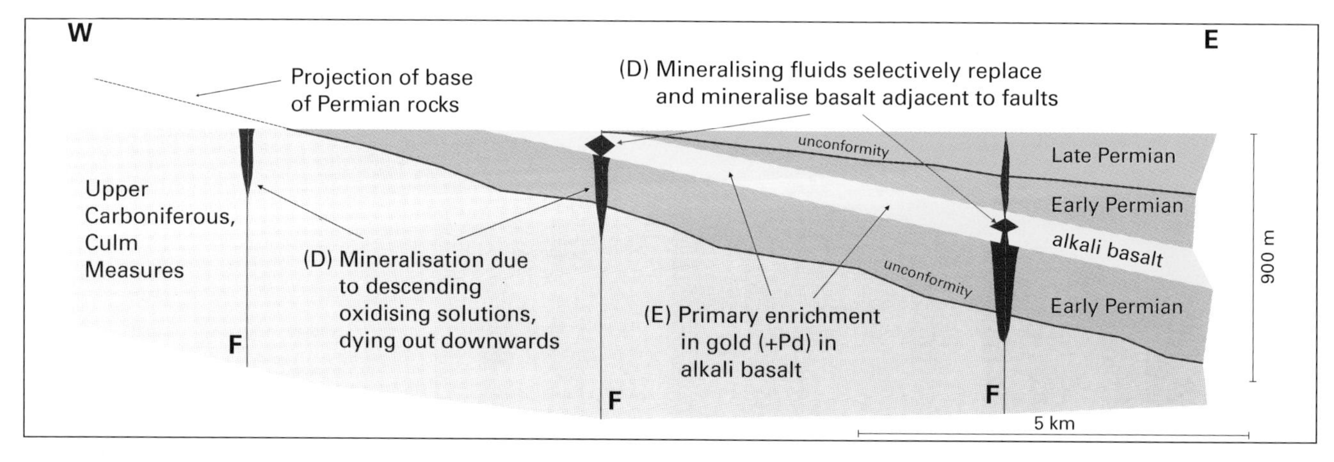

Figure 9 Model for unconformity-related gold mineralisation in Permo-Triassic red-beds.

tion model was developed to account for the association of gold with the boundary between the red-beds and the underlying or adjacent more reduced strata. This model envisages the leaching of gold from a dispersed large-volume source by circulating saline oxidising fluids, transport as a chloride solution, and precipitation of gold on contact with a more reduced system. Thus the contact zone between the red-beds and the underlying or adjacent rocks is the most favourable area for gold deposition (Figure 9). Reconnaissance surveys indicated that the presence of Permian volcanic rocks and evidence of deep weathering in the underlying strata were favourable factors and that two areas with potential for economically attractive deposits were the Crediton Trough (Figure 3) in south-west England and the Mauchline Basin in south-west Scotland (MRP 144).

Relevant MRP reports include:

- MRP 98, 112, 121, 133, 134 and 144.

5.4 Volcanogenic mineralisation

5.4.1 Volcanogenic Massive Sulphide (VMS) deposits

The largest VMS deposit known in Britain occurs at Parys Mountain on the island of Anglesey (Figure 3). It was discovered in 1768 (though it may have been worked in Roman, or even pre-historic times) and cumulative production exceeds 130 000 t of copper from underground workings and open pits (Figure 10). The main period of production was during the late eighteenth and early nineteenth centuries. Modern exploration started in 1955 and since then a number of companies have drilled over 60 km in about 300 surface and underground holes. Originally described as a vein-like structure, it was reinterpreted as Ordovician volcanogenic mineralisation of Kuroko-style by Pointon and Ixer (1980). The deposit is hosted in hydrothermally altered acid and basic volcanics and sediments of Ordovician and early Silurian age (Tennant and Steed, 1997). Massive bedded sulphides and disseminated 'feeder' veins and stockwork mineralisation occur over a zone 2 km long and up to 500 m wide. The property is currently under active exploration by Anglesey Mining plc who sank a 300 m deep exploration shaft at the western end of the deposit (Morfa Ddu) in 1990 to investigate drill-indicated reserves of 4.8 Mt at 1.5% Cu, 3.0% Pb, 6.0% Zn, 57 g/t Ag and 0.4 g/t Au. About 1 km of underground development and a large programme of underground drilling was carried out while a small pilot plant proved the amenability of the ore to conventional processing. Further drilling has since been carried out and the mineralised zones are still open to the north and east. Additional resources of over 30 Mt grading 0.76% Cu were indicated by earlier exploration to the north-east of the shaft site. MRP exploration in north-west Anglesey, including drilling, has indicated the potential for further polymetallic base and precious metal deposits in the area (MRP 99).

A Kuroko-style massive pyrite deposit of Ordovician age at Cae Coch, north-east of Snowdon (Figure 28), may be the distal expression of an undiscovered buried proximal base-metal sulphide deposit (Ball and Bland, 1985). Bimodal volcanic successions of Ordovician age erupted in submarine environments elsewhere in Wales have some potential for VMS deposits. In southern Snowdonia, the MRP carried out an airborne magnetic and EM survey over part of the Harlech Dome and followed this up by detailed ground investigations of twenty six separate areas containing anomalies (MRP 29). This was followed by a regional stream-sediment and panned-concentrate survey with analyses for a wide range number of elements (MRP 74). The surveys revealed a large number of stratabound and VMS mineralisation targets, many of which were not fully investigated. Follow-up MRP investigations in one target area, Benglog, showed barium enrichment in Ordovician sedimentary rocks associated with subaerial to submarine bimodal volcanism but no base-metal mineralisation (MRP 63).

MRP investigations around Treffgarne in south-west Wales (Figure 24), an area with no known economic mineralisation, identified a zone of intense hydrothermal alteration with baryte and pyrite mineralisation in the Ordovician Roch Rhyolite Group of acid volcanics, tuffs and volcanic breccias. There is potential for both VMS and epithermal gold mineralisation in the area and other Lower Palaeozoic volcanic sequences in south-west Wales locally show signs of hydrothermal alteration, but so far little base-metal mineralisation has been found (MRP 86 and 137).

The Middle to Upper Dalradian Tayvallich Formation is a thick (up to 5 km) series of submarine basic tholeitic lavas and high level intrusions associated with the development of the Dalradian extensional basin in western Scotland (Leake, 1982). The formation has potential for base and precious metal mineralisation in a number of environments and deposit styles, including the volcanogenic massive sulphide style described in the stratal aquifer model of Lydon (1988). Other areas with exploration potential in Scotland include Vidlin, in the Dalradian of Shetland where drilling by the MRP and a company revealed stratabound sulphides reaching 10 m in thickness and grading up to 1.19% Cu and 1.27% Zn. The mineralisation is similar in style to several small deposits mined in Scandinavia.

Figure 10 Open pit at Parys Mountain showing extensive alteration of the host rocks.

Stratabound massive sulphides in volcanic rocks are also reported from the Tyrone Igneous Complex in Northern Ireland, which also contains gold and base-metal mineralisation of several other types (Clifford et al., 1992).

5.4.2 Besshi-style deposits

At Gairloch in north-west Scotland (Figure 3) a sub-economic Besshi-style Cu-Zn deposit occurs in deformed metamorphosed supracrustal Lewisian rocks of the Loch Maree Group (Jones et al., 1987; MEG 179). The mineralisation comprises stratiform pyrite, pyrrhotite, chalcopyrite and sphalerite in a 4 m thick quartz-carbonate schist unit with a strike length of at least 1 km. Another 6 m thick barren pyrite-pyrrhotite horizon occurs nearby. Areas of Lewisian metasediments, similar to those hosting the Gairloch deposit, occur on the north side of Loch Maree and elsewhere in north-west Scotland. Exploration in the Glenelg area, 50 km south of Gairloch, found an extensive strike length of an iron-rich chemical exhalite horizon, but without base-metal enrichment (MRP 140). Regional exploration of the whole of the Lewisian outcrop clarified the complex stratigraphy of the Loch Maree and Flowerdale areas and allowed them to be compared to the succession in the Gairloch area (MRP 146). An 8 km strike length of rocks, similar to those in the Gairloch area but partly concealed below shallow Torridonian sedimentary cover, was identified as having potential for the discovery of a Besshi-style Cu-Zn-Au deposit similar to that at Gairloch. Sporadic outcrops of banded oxide and sulphide facies contain significant gold values of up to 4 g/t (MRP 146). A very small outcrop of massive sulphide in Dalradian metabasic volcanic rocks at Garths Ness on the southern tip of Shetland was investigated by Grenmore Holdings but no extensions were found (MEG 249).

5.4.3 Caldera-related deposits

Mineralisation in northern Snowdonia occurs mainly as small veins near the contact of the Ordovician Lower Rhyolitic Tuff Formation and the overlying basic Bedded Pyroclastic Formation but extends down to the underlying sediments. It is related to the development of a resurgent caldera to the south of Snowdon and is considered to be of late-stage fumerolic origin as it predates the Caledonian deformation (Reedman et al., 1985). The mineralisation consists of veins and stockworks of Cu-Pb-Zn sulphides with pyrite and pyrrhotite in quartz. The veins are associated with intense wallrock alteration (Ball and Colman, 1998). Earlier quartz- magnetite breccia veins, with up to 0.1% Sn-W, and later calcite-marcasite-sphalerite veins, also occur. Recent work in the Lake District suggests that at least some early phase mineralisation in the Borrowdale Volcanic Group may also be related to caldera collapse (Millward et al., 1999). Numerous MEG-supported projects have been carried out in Snowdonia, especially by Noranda-Kerr, and all data are available on Open File. However, at least in part due to environmental constraints, relatively little detailed exploration, especially drilling, has been carried out in the Lake District or Snowdonia since the 1970s.

Relevant MRP and MEG reports include:

- Lewisian: MRP 140, 146; MEG 179 (Gairloch).
- Dalradian MRP 4 (Vidlin); MEG 150, 249.
- North Wales (Snowdonia): MRP 22, 29, 63, 74; MEG 75, 76, 77, 78, 79, 80, 87, 88, 92.
- North Wales (Anglesey): MRP 51, 99 and 112.
- South Wales: MRP 86 and 137.

5.5 Carbonate-hosted mineralisation

5.5.1 Irish-style

The main characteristic of Irish-style Zn-Pb mineralisation is syn-diagenetic (but increasingly now thought to be post-lithification) deposition of sulphides in shallow-water carbonate facies, adjacent to major listric faults, at early Dinantian sedimentary basin margins. Mineralisation is generally zinc dominant and its origin is controversial. It has features in common with both the Mississippi Valley Type (MVT) and the SEDEX (SEDimentary EXhalative) styles of mineral deposit which have been described by Anderson and Macqueen (1982) and Large (1983) respectively. An authoritative comparative review of the two styles is provided by Sangster (1990), who considers that their morphological characteristics are dissimilar but that both deposit styles are derived from fluids originating in sedimentary basins. Recent overviews of Irish Zn-Pb mineralisation are provided by Andrew et al. (1986), McArdle (1990) and Hitzman (1995). In recent years the general concensus, backed by new information from the deposits (Galmoy and Lisheen) along the Rathdowney Trend in Cos. Tipperary and Kilkenny, is moving towards an entirely replacement origin for most of the deposits. This envisages sulphide mineralisation replacing a precursor regional dolomitisation of the original limestone host rock (Hitzman, 1995).

Those features in common with the MVT style are largely textural, because of the carbonate host rock, while the features in common with the SEDEX style are largely spatial. A typical MVT deposit occurs at Harberton Bridge where vertical breccia bodies cut Courceyan to Arundian carbonates (Emo, 1986). Lydon (1986) considers that mineralising fluids were generated by dewatering of Devonian to early Carboniferous basins. The fluids reacted with the clastic basin infill at depths of 1–2 km to produce saline, metalliferous brines. A cover of early Carboniferous shales provided thermal insulation to enable the brines to reach temperatures exceeding 200°C under an enhanced geothermal gradient. Periodic extension of the basin permitted release of the metal-bearing fluids up listric faults into the unlithified Carboniferous sediments. Russell (1986) preferred a model in which mineralising fluids were generated by Carboniferous sea water penetrating, and reacting with, the underlying Caledonian basement to increasing depths (up to 10 km) under a normal geothermal gradient during basin extension before being discharged up listric faults. These syn to early diagenetic models are being replaced by post lithification models, though no, single unifying theory is yet accepted for all deposits. For example, Hitzmann (1995) proposed that the Devonian/Carboniferous Munster basin, to the south of Ireland, was responsible for generating the mineralising fluids with a tectonic fluid drive to the north during the Variscan Orogeny, in a similar way to that proposed by Leach and Rowan (1986) and modelled by Appold and Garven (1999) for the main MVT type area mineralisation.

Areas in Britain considered prospective for this type of mineralisation all contain early Carboniferous sediments in extensional, half-graben basins bounded by syn-depositional faults. The geological settings are somewhat similar to those of the Irish deposits, though the extensive Waulsortian reef complex of Ireland is represented only by sporadic reef knolls in the Craven Basin. They include the East Midlands (Plant and Jones, 1989), Craven (Gawthorpe, 1987), Stainmore and Solway (Chadwick et al., 1995) basins (Figure 28). A study in Northern England in the early 1990s of the geological, tectonic, geophysical and metallogenic factors likely to be involved in Irish- and Pennine-style mineralisation (Section 5.5.2) showed that there are a number of areas where early Carboniferous (Tournaisian-Chadian) rocks are likely to be at a depth of less than 500 m below OD (Plant and Jones, 1999). Some of these areas are adjacent to major listric faults, such as the North Solway Fault, and are thus potential sites for Irish-style mineralisation (Figure 11). Extensive areas of Dinantian subcrop, of potential interest for Pennine-style vein and replacement mineralisation, were also delineated during the project. The Craven Basin is more heavily tectonised, and carbonate debris-flows and stratabound Zn-Pb mineralisation occur near-surface in the Cow Ark to Bowland area in the north-west part of the basin. This indicates that suitable depositional and tectonic environments for Irish-style mineralisation could occur at reasonable depths. The MRP located outcropping stratabound mineralisation at Cow Ark and drilled several shallow holes, intersecting encouraging, though uneconomic, Zn-Pb values with up to 3.6% Zn over 2 m. This encouraged BP Minerals to carry out extensive exploration in the area and drill numerous holes in the Marl Hill Moor and Brennand areas. A number of low-grade (~2% Zn+Pb over widths of up to 8 m) intersections of stratiform disseminated style were made in debris beds of the Pendleside Limestone Formation and in stratabound veinlet style in several shale and limestone units. Large amounts of data, including drill core, collected during this work are now available (MEG 241). The basin has also been the subject of more limited investigations by other companies.

Mineralisation along the northern margin of the Lake District (forming the southern margin of the Solway Basin) in rocks of Carboniferous and Ordovician age, is believed on the basis of stable isotope studies to be mainly derived from Lower Carboniferous anhydrite, following the interesection of a thick evaporite sequence in an oil exploration borehole on the north-east margin of the Solway Basin (Crowley et al., 1997). The mineralisation mostly comprises fracture and joint fillings of baryte, with carbonates and base-metals, but baryte is found locally in Carboniferous sandstones. It has been suggested that deposition results from Upper Carboniferous to Lower Permian brine movement along faults at the southern margin of the Solway Basin (Young et al., 1992). If this is true, there would appear to be potential for sub-surface Irish-style mineralisation, particularly as there is evidence for Lower Carboniferous volcanism in the area (Cockermouth lavas). The MRP has investigated a number of areas on both sides of the Solway Basin and has also carried out reconnaissance surveys of the Northumberland Trough. Minor replacement and veinlet Zn-Pb mineralisation was found in several areas. A number of shallow drill holes were completed in the Langholm area on the north side of the basin (MRP 17).

Relevant MRP and MEG reports include:

- Solway and Northumberland basins: MRP 14, 17, 62, 77, 118, 122 and Data Releases 17, 18, 20, 21, 22 and 23; MEG 130 and 229.
- Craven Basin: MRP 66; MEG 241.
- South Pennines: MRP 139.

5.5.2 *Pennine-style*

Dunham (1983) considers Pennine-style mineralisation to be a fluoritic subtype of the Mississippi Valley Type. It generally occurs as fracture-hosted mineralisation in late Dinantian to early Namurian platform carbonates adjacent to Carboniferous shale-dominated basins. The mineralisation comprises large numbers of long (up to several km), narrow (less than 10 m), steeply-dipping oreshoots of limited vertical extent confined to a small number of massive limestone or sandstone beds. Fluorite and galena are the main ore minerals with subsidiary, but locally important,

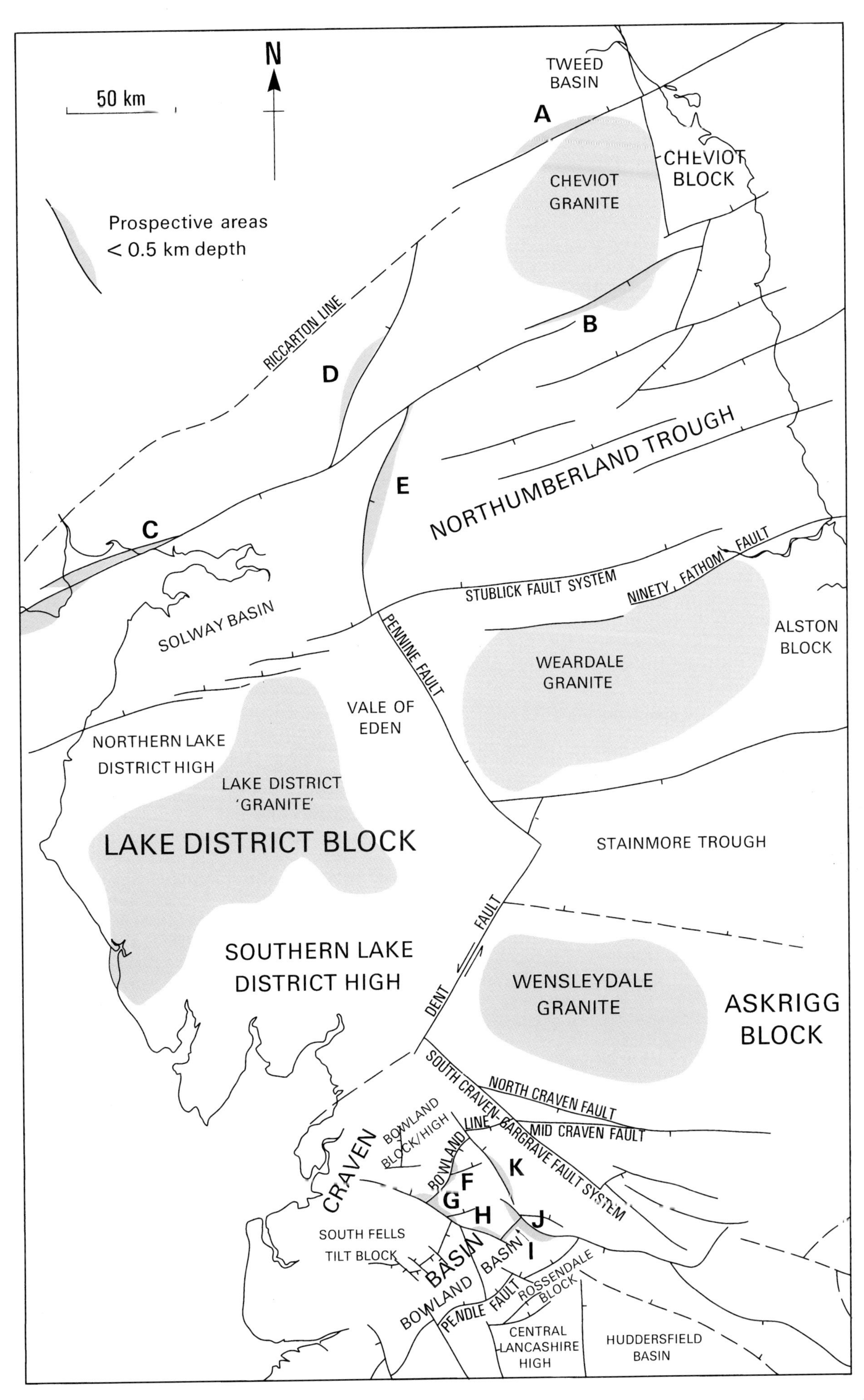

Figure 11 Prospective zones for Irish-style mineralisation based on interpreted zones of shallow (<500m depth) Tournaisian-Chadian rocks adjacent to major syn-depositional faults.

baryte, calcite, sphalerite, witherite, chalcopyrite and quartz. Mineralisation probably occurred during early Permian times. There are a number of significant replacement orebodies, some of which were only partly investigated and/or extracted due to the technologies of the time and the concentration on richer and more easily worked vein deposits. The major Masson (Dunham, 1952) and Dirtlow Rake (Butcher and Hedges, 1987) replacement deposits in the Southern Pennine Orefield have both yielded several hundred thousand tonnes of fluorite.

Two investigations of the Northern (Hunting Geology and Geophysics, 1983; Plant and Jones, 1999) and one of the Southern Pennine (Plant and Jones, 1989) areas, using combinations of geological, geochemical and geophysical digital datasets have indicated potentially prospective areas for Pennine-style mineralisation at depths of less than 500 m below OD outside those where mineralisation has been worked (Figure 11). In addition there are numerous MRP and MEG reports describing local targets and regional surveys which cover most of the Pennine regions and some of the other areas of Dinantian limestone outcrop, such as North Wales, the northern Lake District and the Mendips.

Relevant MRP and MEG reports include:

- North Pennines: MRP 24, 31, 47, 54, 57, 64, 65, 71, 97; MEG 38, 51, 52, 53, 61, 69, 70, 72, 112, 117, 131, 166, 217 and 236.
- South Pennines: MRP 56; MEG 66, 67, 83, 84 and 124.
- North Wales: MRP 75; MEG 105 and 186.

5.6 Mafic intrusion-hosted mineralisation

There are several ophiolite sequences in Britain (Figure 29); Unst in Shetland, Ballantrae in south-west Scotland, Anglesey in North Wales, the Lizard in Cornwall and Tyrone in Northern Ireland.

The Lower Palaeozoic Unst complex has been investigated by the BGS and mining companies for copper-nickel (MEG 136), chromite (MRP 35) and precious metals (MRP 73). Pritchard and Lord (1988), Lord et al. (1994) and Lord and Pritchard (1997) provide additional information on its geological setting. The complex consists of harzburgites, dunites, pyroxenites and gabbros. The ultrabasic rocks have been extensively affected by several phases of hydrothermal alteration. Initial surface exploration (Gunn, 1989) has shown two distinct types of PGE enrichment. The first type is a Ru-Ir-Os dominant assemblage typical of that associated with ophiolitic rocks; the second is a Pd-Pt dominant type, with enrichment of Au, Ni, Cu, As, Sb and Te, which has a PGE distribution similar to that in major layered complexes such as the Bushveld and Stillwater. This second type is thought to have a hydrothermal origin (MRP 73). Additional sampling was carried out by Flight et al. (1994). Leicester Diamond Mines Ltd. carried out reconnaissance drilling in 1999 to test the extent of the PGE enrichment in part of the complex. The only significant intersection is reported as 5.82 g/t Pt and 16.99 g/t Pd over 1.83 m.

In southern Scotland, the Ordovician Ballantrae complex of spilitic pillow lavas and cherts with gabbro and serpentinite contains some chromite horizons (Stone et al., 1986). Drilling proved minor intersections of nickeliferous marcasite (MEG 103). Chromium-bearing serpentinites also occur along the Highland Boundary Fault near Helensburgh (MRP 61).

The Lizard complex is composed mainly of serpentinite with gabbro and metamorphosed basic rocks. The BGS has carried out soil sampling (Smith and Leake, 1984) and drilling (Leake and Styles, 1984) in this poorly exposed area, in part to aid geological mapping. Titanium enrichment, with enhanced vanadium was found in a number of boreholes with up to 7.6% TiO_2 over 2.3 m and up to 5% V_2O_3. However, the distribution of the oxide-rich fraction suggests that the potential for large volumes of Fe-Ti-rich material is limited (MRP 117). The small Polyphant ultrabasic intrusion near Launceston has also been drilled (MRP 32), but no significant mineralisation was encountered.

In northern Anglesey there are several small poorly-exposed serpentinite intrusions, hornblende gabbros, spilitic rocks and cherts in a complex stratigraphic and tectonic setting which is still the subject of debate. In the south-east of the island gravity and magnetic anomalies occur over poorly exposed hornblende schists providing little surface evidence of any mineralisation. Geochemistry indicates that the schists were originally tholeiitic ocean basalts, while the geophysical evidence suggests that a sub-surface layered basic igneous complex may be present (MRP 127). Small layered basic intrusions containing sub-economic Ti concentrations occur in the Lleyn Peninsula in North Wales (MEG 81) and at Carrock Fell in the Lake District.

Remnants of ultrabasic and basic layered complexes within an Archaean (Lewisian) high-grade terrain occur in north-west Scotland (Windley, 1982). No economic mineralisation is known but several drillholes near Scourie cut massive and disseminated pyrite-pyrrhotite lenses (MEG 2).

The major Caledonian basic and ultrabasic intrusive complexes of north-east Scotland (Wadsworth, 1982) have been investigated for copper, nickel, titanium and PGE by several mining companies and the BGS. There are numerous MEG Open File reports on this work with details of reconnaissance stream-sediment geochemistry, soil sampling, rock sampling, geophysics (magnetics and EM), and about 15 km of core drilling. Some core is held by the BGS. An airborne EM and magnetic survey of the whole area was flown in the early 1970s and the aeromagnetic data have now been converted to digital form (MRP 136). Some of the masses, such as those at Huntly and Belhelvie, include significant ultrabasic and olivine cumulate rocks. Others such as Insch and Cabrach-Morvern are mainly basic and are enriched in iron, titanium and phosphorus compared to average values for these rocks. The rocks have been tectonised in a series of regional shear zones as described by Munro (1986).

The most extensive exploration programme was carried out by Exploration Ventures Ltd (a joint venture between Consolidated Gold Fields and Rio Tinto-Zinc). Low-grade (less than 2%) Cu-Ni sulphide mineralisation in norites and gabbros contaminated by pyritic and graphitic Dalradian metasedimentary rocks was located by the exploration work. Two areas with potentially economic mineralisation are at Littlemill, near Huntly, where a drill-indicated resource of 2 Mt at 0.52% Ni and 0.27% Cu has been outlined, and Arthrath, north of Ellon, with a resource of 17 Mt at 0.21% Ni and 0.14% Cu (Fletcher et al., 1997). The presence of widespread pyrite and graphite, together with the faulted nature of the mineralisation, gave rise to great problems in interpreting geophysical data and in correlating drill intersections.

Noranda-Kerr Ltd found anomalous copper and nickel values in deep overburden samples on the shear zone trend at the eastern end of the Insch intrusion near Old Meldrum which were followed up by the MRP (MEG 95 and MRP 119). Anomalous barium and zirconium values occur in syenites associated with the Insch gabbro. The area still has potential for the discovery of mineralisation but the complexes are generally of low relief with thick glacial deposits which hamper exploration. Deep overburden sampling, coupled with detailed magnetic surveys, is probably the best exploration technique. Parts of the Insch and Belhelvie intrusions have been investigated by Aberdeen University using magnetics and shallow bedrock drilling (Ashcroft and Boyd, 1976; Ashcroft and Munro, 1978; MRP 124).

Additional exploration has been carried out for platinum-group elements associated with deformed ultramafic rocks near Huntly (MRP 115 and 124). Drilling near Kelman Hill south-west of Huntly showed that a PGE-enriched serpentinite contained over 70 ppb Pt + Pd over 6.5 m with a maximum of 280 ppb. Similar concentrations were found in the Succoth–Brown Hill area (MRP 115). Exploration in the area north-west of Huntly found some enrichment in PGE, including up to 584 ppb Pt and 93 ppb Pd in a mineralised pegmatitic pyroxenite from the Bin Quarry south of Littlemill (Fletcher and Rice 1989). Follow-up drilling showed that these bodies are narrow and not persistent.

Relevant MRP and MEG reports include:

- Scotland: MRP 35, 61, 73, 115, 119, 124; MEG 2, 7–37 inclusive, 41, 91, 95, 96, 97, 107–111, 103 and 136.
- South-west England: MRP 32 and 117.
- Wales: MRP 127.

5.7 Epithermal gold mineralisation

Low-sulphidation epithermal gold mineralisation occurs at Rhynie (Figure 24) in the north-east Grampian Highlands in an outlier of Lower Devonian sedimentary and volcanic rocks overlying Dalradian Southern Highland Group strata. Geochemical, alteration and textural features indicate an origin in a hot-spring setting related to late Caledonian volcanism. The altered rocks and chert sinters contain high levels of Au, As and Sb and are locally enriched in W, Mo and Hg (Rice et al., 1995). Textural and geochemical evidence for similar epithermal mineralisation in the headwaters of the River Brora in Sutherland has been reported by Crummy et al., (1997).

In western Scotland the Lorn plateau is formed of Devonian andesitic lavas and tuffs, which are contemporaneous with the Lagalochan intrusion and its associated copper-gold mineralisation (Section 5.10), and unconformably overlie Dalradian schists. It is thought that the exposed Lagalochan complex could have lost 1000 m by erosion and that only the basal sections of the mineralisation are preserved (Harris et al., 1988). The Devonian lavas may therefore conceal, at relatively shallow depth, preserved high level epithermal mineralisation associated with Caledonian intrusions. This zone is envisaged as similar to that described by Panteleyev (1986) from Canadian epithermal precious metal vein deposits. They form near the surface, in upward branching veins associated with extensional tectonics, in subaerial volcanic rocks which are commonly intermediate in composition.

MRP surveys identified widespread alluvial gold in the Ochils Hills of central Scotland. The highest concentrations are found in Borland Glen, where Lower Devonian andesitic lavas and pyroclastic rocks are intruded by a dioritic body and porphyry dykes. Locally intense argillic hydrothermal alteration and brecciation are indicative of an epithermal setting. Follow-up investigations by BGS and a company in the early 1990s failed to identify gold mineralisation in bedrock. Some alluvial gold has been extracted but the potential of the source remains to be defined.

Other Devonian rocks in northern Britain have potential for this style of mineralisation. In the Cheviot Hills several areas of hydrothermal alteration and minor mineralisation were discovered in Lower Devonian andesitic lavas and a high-level intrusive porphyry complex (MRP 91). Only a small number of samples were analysed for gold as this was not the target of the survey, but the geology shows features in common with epithermal precious-metal deposits described by Sillitoe (1993).

In Northern Ireland evidence of this type of mineralisation is found with other styles in the Tyrone igneous complex, notably as disseminated gold and base-metal mineralisation in a rhyolite dome at Cashel Rock (Figure 3), 12 km due south of the Curraghinalt gold deposit (Clifford et al., 1992). The mineralisation is erratic, but grades of up to 30.51 g/t Au over 3.63 m have been intersected by scout drilling.

Relevant MRP and MEG reports:

- MRP 91 and 116.

5.8 Gemstones

A wide range of gem material has been recorded in the rocks of Britain, particularly those in Scotland and south-west England. In addition to sapphire, ruby and possible diamond, the country has yielded topaz, beryl, cairngorm and amethyst, together with gem-quality garnet, tourmaline, agate and zircon. There is even an unconfirmed report of Scottish emerald but, except for local industries based on some semi-precious stones, there has been no commercial production. Stimulated by diamond discoveries in northern Canada companies have been re-appraising the potential of northern Europe for gemstones since the mid-1990s. This led to commercial surveys in Ireland, which have established the presence of hitherto-unrecorded alluvial sapphires in Co Donegal and kimberlite indicator minerals in Co Fermanagh, close to where an alluvial diamond was discovered in the nineteenth century.

The geological continuity between Ireland and Britain raised interest in the gemstone potential of Scotland, which was also stimulated by the discovery of sapphire megacrysts ranging from 10 to 30 mm in a xenolithic camptonite dyke at Loch Roag on the Isle of Lewis, Outer Hebrides (Figure 3). A 9.6 carat cut stone from this locality was valued at £60–70 000 in 1995.

A recent study by the Mineral Reconnaissance Programme (MRP 135) evaluated the diamond potential of Great Britain and identified the Lewisian Terrane of north-west Scotland as the most prospective area, because it is a fragment of the ancient Laurentian craton which seismic data indicates has crust thick enough to have induced the crystallisation of diamond in the upper mantle. No kimberlites are known in this area, but exposure is very poor over large areas and existing aeromagnetic data are too widely spaced to give an adequate indication of whether any diatremes are present. The nature of the basement under the rest of the Scottish Highlands is less clear, but it could be partly of Lower Proterozoic age and have some potential for the occurrence of diamonds in lamproitic rocks. On the basis that diamonds could crystallise within a cold subducted slab which caused local depression of the geothermal gradient and then be tapped by lamprophyres, alkali basalts or nephelinites, some potential also exists in the Southern Uplands of Scotland and south-west England, where such rocks are comparatively widespread and locally contain evidence of mantle origin. Prime targets within the suite are monchiquites carrying mantle xenoliths. Diamonds

could also occur in palaeoplacers within the Proterozoic Torridonian sedimentary rocks of north-west Scotland.
Relevant MRP and MEG reports:

- MRP 135.

5.9 Granite-related mineralisation

The Variscan granite province of south-west England is a major mineralised region historically famous for its tin deposits and early descriptions of zoned granite-related mineralisation. Tin, tungsten, copper, lead, zinc and uranium have all been extracted and targets of interest remain. Detailed data are provided in numerous MRP and MEG reports. A comprehensive, modern summary is provided by Alderton (1993) and additional information on the mineralisation is given in Dominey et al. (1996). The main targets have been stockworks or sheeted vein systems, such as those at Hemerdon and Redmoor (Figure 3). At the Hemerdon W-Sn deposit near Plymouth, where sheeted veins occur at the apex of a small granite cusp, ore reserves of 42 Mt grading 0.18% WO_3 and 0.02% Sn have been identified (MEG 187). Christoffersen and King (1988) describe the evaluation of this deposit, including exploration drilling, metallurgical testwork and ore reserve calculation. Planning permission has been granted for mining. Other tin-tungsten prospects include Cligga Head, a greisenised granite on the north Cornwall coast which has been examined by adits and drilling, and Redmoor, where drilling has shown a large, low-grade sheeted vein swarm up to 80 m wide and several hundred metres long (Newall and Newall, 1989; MEG 165). Drilling in an area north of the St Austell Granite has proved several high-grade veins and flat-lying zones in the killas (MEG 237). The most comprehensive available data on exploration for tin-tungsten are around the Mulberry open pit near Bodmin, a large low-grade deposit. Exploration by the BGS and mining companies has shown tin values to extend beyond and below the open pit. Drilling carried out to prove an underlying granite cusp at shallow (<500 m) depth was unsuccessful and some of the higher-grade intersections of vein mineralisation, up to 11 m at 1% Sn, contained refractory tin silicates (MRP 48; MEG 73 and 198).

The uranium potential of south-west England has also attracted attention and many occurrences have been found (Bowie et al., 1973; MRP 45 and 110). Uranium was worked from veins at a few localities in the early part of the century, but no known occurrence is of sufficient size to merit working under current economic conditions.

Granite-related structurally-controlled uranium mineralisation also occurs in Scotland associated with end-Caledonian granites in the Highlands (e.g. Helmsdale, Bennechie, Hill of Fare), and with the Devonian Criffel Granodiorite near Dalbeattie in the south-west (Rice, 1993), but to date only small sub-economic concentrations have been located.

Vein-style granite-related mineralisation is associated with components of the Lake District batholith, notably tungsten mineralisation at the margin of the Skiddaw Granite which was exploited from the Carrock Fell Mine. Low-grade copper, molybdenum and bismuth mineralisation in the Shap Granite was investigated by a company and the MRP as a porphyry-style target (MRP 109). More recently evidence of further tungsten and tin mineralisation associated with the batholith has been found close to its southern margin in the Black Combe area (MRP 128). Many of the other Lake District mineral veins that are hosted in the Ordovician Borrowdale Volcanic Group and underlying clastic sedimentary rocks of the Skiddaw Group are probably sourced from the volcano-sedimentary succession, but may be genetically related to the batholith in that it provided the energy to drive convecting cells bearing mineralising fluids (Stanley and Vaughan, 1982; Cooper et al., 1988).

Relevant MRP and MEG reports include:

- MRP 1, 2, 11, 25, 34, 44, 45, 48, 49, 81, 82, 83, 95, 107, 109, 110, 128; MEG 1, 65, 73, 106, 119, 129, 159, 162, 163, 164, 165, 182, 187, 188, 191, 195, 198, 199, 203, 208, 211, 212, 218, 237, 251, 256 and 264.

5.10 Calc-alkaline porphyry-style mineralisation

The search for this type of mineral deposit in Britain was triggered in part by the discovery of the Coed-y-Brenin copper deposit in North Wales (MEG 5; Rice and Sharp, 1976) in 1968. 110 holes were drilled to depths of up to 300 m to outline a resource of about 200 Mt at 0.3% Cu with minor gold values. Low-grade concentrations of copper, molybdenum, tin, tungsten and uranium were found in association with a number of mineralised Caledonian calc-alkaline porphyritic granites in Scotland, northern England and Wales. Later, areas of extensive hydrothermal alteration in high-level intrusives of the west and central Scottish granites were investigated for gold mineralisation especially in or near shear zones. Targets included Lagalochan and Foreburn.

Phelps Dodge Corporation NL, and later Consolidated Gold Fields Ltd, carried out regional exploration, mainly using stream-sediment geochemistry with limited follow-up, over most of the Scottish Caledonide granitoids in the 1970s. No discoveries were reported (MEG 2 and 3) but later work located quartz veins up to a metre wide, with sporadic molybdenite clusters in a small greisenised granite stock at Chapel of Garioch, west of Inverurie (MRP 100 and MEG 17), and localised disseminations and veins of wolfamite and cassiterite in greisenised granite in Glen Gairn, near Ballater (Webb et al., 1992). The MRP reported low-grade molybdenite mineralisation associated with Caledonian granites west of Lairg (MRP 3).

In the south-west Highlands of Scotland disseminated polymetallic mineralisation of epithermal-porphyry style is found at Lagalochan, associated with the last phase of the Kilmelford calc-alkaline dioritic-granodioritic intrusive suite (MRP 9 and MEG 258). Three phases of base- and precious-metal mineralisation and multiple hydrothermal alteration events were identified. Cu-Mo-Au in veinlets and disseminations was followed by shear-related Pb-Zn-Ag-Au-As-Sb mineralisation and, lastly, Pb-Zn-Ag carbonate veins. Mineralisation of porphyry type has also been recorded in association with high level intrusions elsewhere within the Dalradian belt at Ballachulish (MRP 43), Garbh Achadh (MRP 23), and Tomnadashan (Pattrick, 1984). All are of probable late Caledonian age.

MRP and company investigations around the Foreburn igneous complex, near Ayr in southern Scotland (Figure 3), have shown that alteration, including sericitisation and tourmalinisation, has affected zones within the complex which is composed of diorite, tonalite and feldspar porphyry. Disseminated and vein sulphides are common with up to 1.4 g/t Au especially in shear zones. The complex was thought to have similarities with some Canadian porphyry copper deposits (MRP 55). Subsequent exploration by RioFinex at Foreburn, including drilling, found quartz-tourmaline-sulphide veins and stockworks with significant gold values (Charley et al., 1989). In the same area, the Black Stockarton Moor high level intrusive complex of Caledonian age contains low-grade copper mineralisation of porphyry copper style (MRP 30). There are a number of similar intrusions in southern Scotland, some of which are known to be mineralised, for example Cairngarroch (MRP 39).

MRP investigations in the Llandeloy area (Figure 24) of south-west Wales identified an intermediate dioritic intrusive complex of Cambro-Ordovician age with strong propylytic and, locally, potassic, hydrothermal alteration and low-grade disseminated copper mineralisation. This may reflect erosion to the deep levels of a copper porphyry system, particularly as overlying sediments of possible Tertiary age are enriched in copper (MRP 78).

In none of the bodies investigated, apart from that at Coed-y-Brenin in North Wales, have significant zones of more than 0.3% Cu been identified. The apparent paucity of copper may result from the emplacement of relatively anhydrous granites into tectonised and anhydrous metasediments, especially in the Dalradian and Moinian of Scotland which were highly metamorphosed during Caledonian and earlier orogenic events. Further information is given in Plant et al. (1983), Plant (1986) and Rice (1993). The latter contains a synthesis of information on Caledonian igneous activity and mineralisation and includes useful and comprehensive tables and maps of the numerous mineralised intrusions.

There is an extensive area of Ordovician (Caradoc) volcanic rocks in County Tyrone, Northern Ireland, between Omagh and Cookstown. The volcanic suite includes spilites, andesites and rhyolites with associated black shales, cherts, tuffs and small, high-level intrusions of gabbro, diorite and granite (Stillman, 1982). A review of exploration for base-metal mineralisation in the area is given by Leyshon and Cazelet (1976), who focused on the potential of the intrusions as hosts for porphyry-style mineralisation. Some of the intrusives host significant, but sub-economic, mineralisation of this type with drill intersections of up to 46 m at 0.17% Cu. More recent exploration has located other styles of mineralisation, notably disseminated gold and base-metals of possible epithermal origin (see above) in a rhyolite dome at Cashel Rock, 12 km due south of the Curraghinalt gold deposit (Clifford et al., 1992).

Relevant MRP and MEG reports include:

- Northern Scotland: MRP 3; MEG 2 and 17.
- North-east Scotland: MRP 96, 100 , 120; MEG 17.
- Western Scotland: MRP 9, 23, 43, 76, 80, 94; MEG 98, 226 and 258.
- South-west Scotland: MRP 18, 19, 21, 30, 39, 42, 46, 55, 58; MEG 2, 135, 257, 259, 260 and 261.
- Northern England: MRP 7, 33 and 60.
- Wales: MRP 38, 78; MEG 5.

5.11 Other settings with potential for mineralisation

5.11.1 Alkaline intrusive complexes in north-west Scotland

Major Caledonian alkaline ultrabasic to acid intrusive complexes, with rock types ranging from pyroxenite to syenite, occur around Loch Borralan and Loch Ailsh (Figure 29) in north-west Scotland adjacent to the Moine Thrust. Concentrations of apatite, magnetite, base-metals, and PGEs have all aroused interest (Notholt et al., 1985). A base-metal exploration programme was abandoned at an early stage leaving a significant IP chargeability anomaly, without accompanying magnetic anomaly, untested (MEG 113). The Borralan (MRP Data Release No. 8) and Loch Ailsh (MRP 131) complexes were subsequently investigated for their platinum potential. In the Loch Borralan area samples from 37 holes drilled during previous phosphate and base-metal exploration programmes were analysed for their PGE content and a further four holes drilled. Values of up to 1000 ppb Pt + Pd occur in the more recent holes. In the Loch Ailsh area widespread, low-tenor PGE (100 ppb Pt and Pd) and Cu enrichment was found with higher values to 300 ppb Pt + Pd occurring sporadically in one area. Sperrylite was identified using automated microprobe analysis, in association with a suite of arsenides, antimonides, selenides and tellurides. Resurveying the IP anomaly did not indicate the presence of any significant sulphide mineralisation. Another large alkaline complex occurs at Loch Loyal, 40 km to the north-east (Sutherland, 1982). Local enrichments of REE and Th were found during MRP investigations in the Cnoc nan Cuilean area (MRP Data Release No. 11). Minor Au-Ag-Te mineralisation occurs in the Ratagain alkaline igneous complex in north-west Scotland (Alderton, 1988). The complex is intruded into Lewisian and Moinian metasediments and the mineralisation occurs as quartz-fluorite-calcite-sulphide veins.

Relevant MRP and MEG reports include:

- MRP 131 and Data Releases 5, 8 and 11; MEG 113.

5.11.2 Basin-related vein and replacement style mineralisation in Central Wales

The Central Wales orefield (Figure 28) in Lower Palaeozoic clastic sedimentary rocks, dominantly of turbidite origin, has potential for vein deposits in depth and for stratabound deposits similar to those of the Van 'flats'. These were a series of ore bodies which occurred in a 5 m thick sandstone unit and became zinc-rich towards the east. Some mineralisation remains unworked (Jones, 1922) and the silver content of the ore was high. Mason (1997) published a review of the paragenesis of the ore minerals and reported the discovery of a number of nickel and cobalt minerals. The mineralisation occurred in stages at 390 Ma and between 360 and 330 Ma (Fletcher et al., 1993). A reconnaissance stream-sediment survey has been carried out by the MRP over much of the orefield (MRP 5). The report is accompanied by a 1:100 000 scale geological map showing the location of the numerous abandoned mines. Reviews of the mineral potential of selected deposits in the region are given by Hughes (1988) and Hall (1989). The basin also has potential for mesothermal turbidite-hosted gold deposits (see above). The orefield is being remapped by the BGS and new 1:50 000 maps will become available during the next five years.

Relevant MRP reports include:

- MRP 5 and Data Release 13.

5.11.3 Estuarine and offshore placer deposits

Exploration for placer tin deposits has been carried out in south-west England in several places onshore and off the north and south coasts of Cornwall (see under tin in Appendix 1). Plans to dredge the top metre of sediment off the north Cornwall coast were put forward by Marine Mining following extensive exploration (MEG 64). Exploration drilling for deeper placer deposits at the base of buried offshore river channels was carried out by Billiton UK in Mount's Bay and St Austell Bay (MEG 220). Billiton also carried out a detailed examination of Goss Moor, north of the St Austell Granite, and drilled over 400 shell and auger holes to bedrock (MEG 210). Smaller-scale onshore alluvial exploration projects in Cornwall are reported in MEG 182 and 211. Research into the origins of the alluvial deposits is reported under Tin in Appendix 1.

In the late 1960s RioFinex planned to examine the Mawddach estuary in North Wales for placer gold derived from the Dolgellau gold deposits and more recently the MRP drew attention to the presence here and in some other estuaries in Wales of grey nodular monazite which is characterised by a low thorium content (MRP 130).

There are several areas in Scotland with potential for nearshore or offshore deposits. In the west, the basic intrusions of the Tertiary Volcanic Province are the source of chromite and olivine sands, with other heavy minerals, discovered during MRP investigations off the south-west coast of Rum (MRP 106). Further local concentrations are likely to exist. The large basic masses of north-east Scotland have shed large amounts of magnetite and ilmenite into the North Sea. Minor placer enrichments have been found in coastal dunes at Rattray Head, near Fraserburgh (Colman, 1982) but the offshore potential is unknown. Other areas with potential for offshore deposits include Helmsdale in Caithness (gold) and Skye and Mull in western Scotland (magnetite, ilmenite and chromite). Manganese nodules with over 10% Mn occur in Loch Fyne on the west coast of Scotland.

Relevant MRP and MEG reports include:

- MRP 106, 130.
- MEG 64, 182, 210, 211 and 220.

5.11.4 Breccia pipe deposits

The Glasdir mine, north of Dolgellau, worked copper with minor gold from a breccia pipe deposit close to, and possibly genetically related to, the Coed-y-Brenin porphyry copper deposit (Allen and Easterbrook, 1978). Other pipes have been mapped in the same area but there is no record of them containing mineralisation at surface. Copper mineralisation has also been reported in appinitic breccia pipes hosted by Dalradian metasediments on the Ardsheal Peninsula in the south-west Highlands.

5.11.5 Skarn-type copper deposits

These are relatively rare in Britain, but have been worked on the northern margin of the Dartmoor Granite. Drilling by the MRP to investigate the extent of this mineralisation indicated that it is more widespread than previously anticipated but of lower grade (MRP 101). Gold values to 40 g/t have been recorded in sulphide concentrates from drill core (Beer and Fenning, 1976). There is accompanying As, Bi, W and Sn, but the latter is largely held in garnet. Magnetite-chalcopyrite-bearing skarns also occur on the island of Skye marginal to a granitic intrusion of Tertiary age (Groves, 1952).

5.11.6 Epigenetic copper deposits in limestones

At Ecton Hill in the English Midlands (Figure 23) about 100 000 t of ore containing 15% copper have been extracted from two pipe-like sub-vertical orebodies and associated veins cutting an anticline of interbedded Visean limestone and shale (Critchley, 1979). Chalcopyrite with calcite and baryte occur in at least five cavities over a vertical distance of about 400 m. Other small but very high-grade copper deposits have been mined in cavernous limestones of Carboniferous age elsewhere in Britain. These include deposits near Llandudno in North Wales (Williams, 1993), at Llanymynech in the Welsh Borders and near Middleton Tyas in Yorkshire (MRP 54). These deposits may have formed as a result of the downward migration of copper-rich fluids from the overlying Permo-Triassic (Masheder and Rankin, 1988), although some authors consider adjacent Carboniferous basins to be a more likely source.

6 MINERAL EXPLORATION TECHNIQUES IN BRITAIN

Britain lies between latitudes 50° and 60°N and has a temperate climate. The terrain is generally subdued but in northern and western regions Palaeozoic and older rocks form higher ground with peaks up to 1300 m. Rainfall varies from 500 mm in the south-east to over 2500 mm in the west. Exploration can be undertaken throughout the year on lower ground and in the south, but the higher ground in Scotland and Wales is usually snow covered, cold and often windy in winter. The entire country is served by a dense network of surfaced roads and nowhere, except in some mountainous areas in Scotland, is more than 5 or 10 km from a road. Most of lowland Britain is intensively cultivated; upland areas tend to be used for sheep farming and/or for game (grouse, deer etc). as well as for other recreational activities. Exploration programmes should be planned to take account of shooting and fishing interests.

6.1 Geochemistry

6.1.1 Stream sediment

Drainage patterns are very well developed in the upland areas of northern and western Britain which form the most favourable areas for metalliferous mineral exploration. Stream sediment geochemistry is therefore an excellent exploration method in Britain and has been widely employed by the Mineral Reconnaissance Programme and private sector companies.

Nationwide surveys employing stream-sediment sampling have also been undertaken to provide baseline information for mineral exploration, agricultural and environmental purposes. The first major survey was carried out in 1969 (Wolfson, 1978). This covered England and Wales at a density of 1 sample per 2.5 km^2 for 21 elements. A second survey covered Northern Ireland at a similar density (Applied Geochemistry Research Group, 1973) and revealed an arsenic anomaly over the Sperrin Mountains where the Curraghinalt gold deposit was later found.

The BGS G-BASE (Geochemical Baseline Survey of the Environment) Programme has been conducting a systematic drainage survey of Great Britain since 1972. The survey began in the north of Scotland and is progressing southwards. Sampling is currently being carried out in the English Midlands (Figure 12). The sample density is about 1 sample per km^2, using first or second order streams. The survey uses careful site selection, a large sample volume and standardised analytical techniques to achieve a highly reproducible output (Plant et al., 1984). Three sample types are collected; (i) a <150 µm mesh active stream sediment which is analysed for up to 31 elements, (ii) a panned heavy-mineral concentrate made from <2 mm stream sediment which is examined microscopically in the field for gold and other heavy minerals, but which is not routinely analysed and (iii) one or more water samples (see below). In areas with poor surface drainage soil sampling is employed to provide additional information. Further information on G-BASE products is given in section 8.1.

More detailed stream-sediment surveys were used by the BGS Mineral Reconnaissance Programme (MRP) to define smaller areas worthy of intensive investigation for specific targets. The Programme made extensive use of panned heavy mineral concentrates made from the <2 mm fraction of the active stream sediment. Chemical analysis of the concentrates was commonly compared with sediment and water or float sample analyses from the same sites to help provide additional information on the source of anomalies (Coats et al., 1994). Mineralogical examination to determine the cause of anomalous levels of various elements in concentrates was commonly employed. This enabled contaminants, such as lead shot, to be identified. The use of panned concentrate grains is particularly useful in gold exploration when combined with EM-probe microchemical analysis of gold grains. Work by the BGS et al. in Britain has recognised a number of different groupings of gold-grain chemistry, based on grain zonation and inclusions of sulphides, selenides, alloys and other minerals, which can be related to the type of bedrock mineralisation from which the alluvial gold is derived (Leake et al., 1993, 1997 and 1998).

A novel method of freeze core sampling to obtain a complete stream-sediment sample, without significant loss of fine material — especially gold, is described by Petts et al. (1991).

The interpretation of stream sediment data in Britain may be limited by:

a. Glaciation. Most of Britain was glaciated during the Pleistocene and extensive deposits of boulder clay, lacustrine sediments and fluvioglacial sands and gravels may comprise most of the stream sediment, which may therefore be diluted by exotic material and not reflect local bedrock composition. Care is needed in site selection and data interpretation.

b. Contamination. Much of Britain is densely populated and intensively farmed, and there is a long history of metal mining and other human activities that have affected the distribution and availability of metals. Contamination of streams from multiple sources is therefore common but often erratic with densely populated and lowland areas most affected. The effects of contamination may be obvious, as with lead shot from shooting or fishing producing elevated levels of Pb+Sn+Sb, batteries (Pb+Sb) and brass (Cu+Zn), but sometimes it may be more subtle where agricultural chemicals (for example U in phosphates) have been used on a wide scale or airborne pollutants deposited, raising the local background. Panned concentrates can often be used directly to identify heavy metallic contamination. Cooper and Thornton (1994) provide an extensive overview, including a summary of interpretative methods used to clarify the sources of anomalies and illustrated with several examples from Britain.

c. Hydrous oxide precipitates. Precipitates of iron and manganese minerals are common in many British streams, due to interaction of acidic surface and groundwaters from peaty upland areas and similar environments with more alkaline and oxidising conditions met at surface and on lower ground. Co-precipitation or adsoption of other metals is a frequent cause of elevated metalliferous concentrations. The process and its products can be used for exploration purposes but careful data interpretation is necessary to avoid pursing anomalies with no mineralised source.

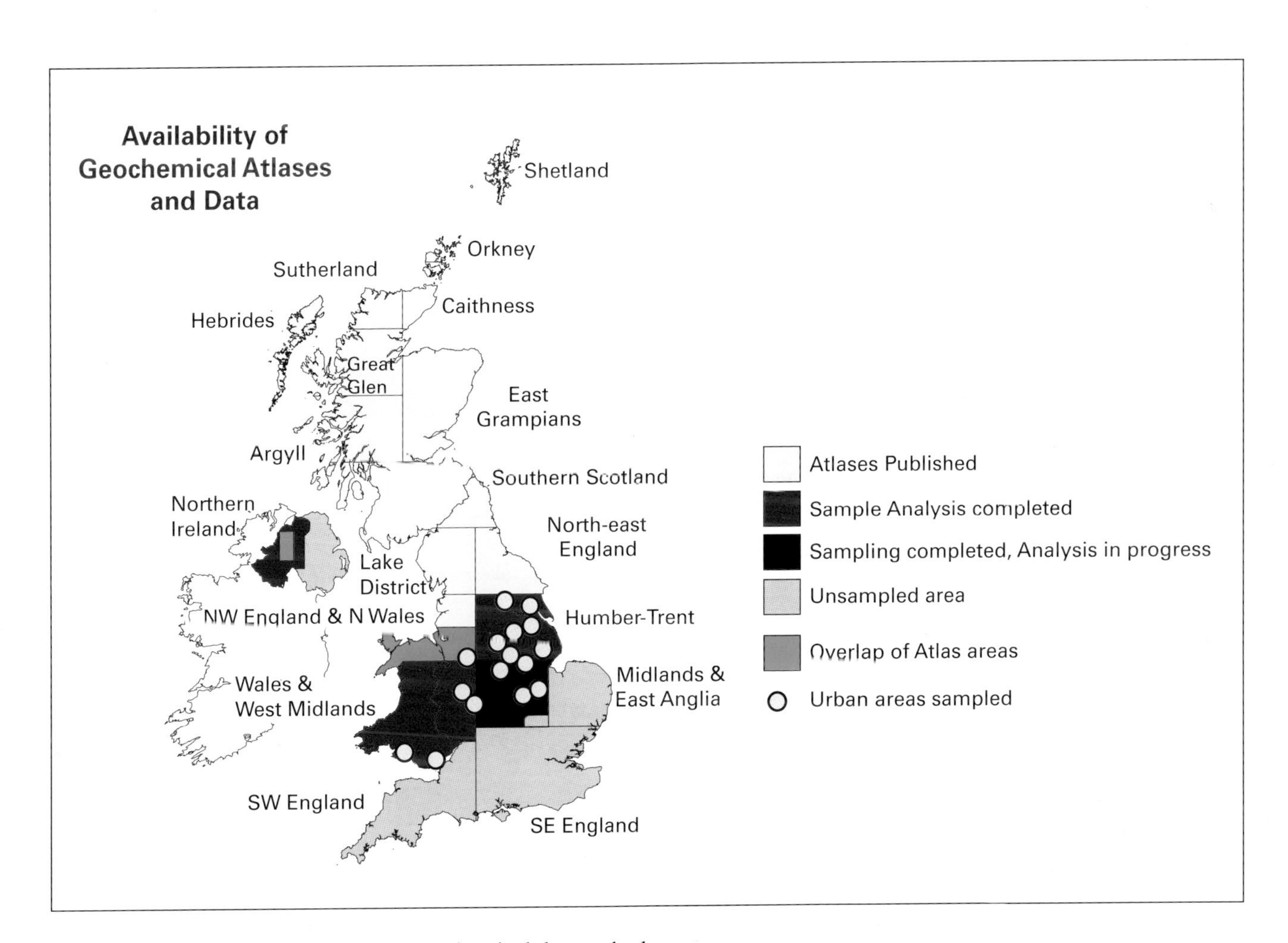

Figure 12 Availability of G-BASE geochemical data and atlases.

A more detailed description of the use and interpretation of stream-sediment geochemistry can be found in the G-BASE regional geochemical atlas series (e.g. British Geological Survey, 1997). Plant et al. (1997) provide an overview of the use of regional geochemistry in understanding the metallogeny of the Caledonides. Panned-concentrate sampling methods are outlined in Leake and Aucott (1973) and Darnley et al. (1995) detail methods of stream-sediment sampling.

6.1.2 Soil

Most soils are composed of weathered bedrock but in much of Britain many are developed on glacial or fluvioglacial deposits. The Soil Survey of Great Britain publishes maps and memoirs which provide information on soil types and conditions. They are produced mainly for the agricultural community and also include some geochemical data. For example, the nickel exploration programme in north-east Scotland was initiated following references to the nickel content of a particular soil type developed over ultrabasic rocks in the Belhelvie area (Glentworth and Muir, 1963). The same source mentioned excess molybdenum in soils which was affecting cattle in the area, and follow-up led to the discovery of a number of molybdenum prospects in the Grampian Region (MEG 17 and MRP 100). Soils with a relatively high background level of copper were recorded from South Wales (Bradley et al., 1978) and, along with geological factors, led to the discovery of porphyry copper mineralisation in the area (MRP 78).

Soil samples of about 200 g are normally collected from the B or C horizons in well-developed profiles using a 1 m hand auger. A-horizon (i.e. organic near surface layers) are rarely sampled for exploration purposes in Britain, at least in part because this horizon is particularly vulnerable to contamination (see below). Base-of-overburden samples are the preferred sample type where glacial deposits are widespread or thick, or there is reason to suspect transport of elements of interest. The BGS and company prospectors commonly use light overburden drills (Cobra, Wacker or Minuteman) for this purpose while a hand auger is normally used for sampling the B and C horizon at depths of up to a metre. A fine size fraction, typically between <40 and <150 μm, depending on soil type and target, is usually taken from the dried and disaggregated sample for analysis. Samples are typically collected along traverse lines at a spacing related to the size of target and predicted dispersion halo, generally 10 to 50 m apart.

The interpretation of soil geochemical data may be constrained in Britain by the same processes that affect drainage sampling:

a Glaciation and associated events. Glacial processes can result in thick exotic deposits which have no relation to bedrock geology, and their contained clays may prevent upward migration of metal ions. However, the presence of large erratics or boulder trains does not necessarily mean that the underlying fine till as also exotic. In south-west England a number of strand lines are developed corresponding to different sea levels which fluctuated up to 130 m above present sea level (Camm and Hosking, 1985). These can commonly carry high levels of resistate minerals including cassiterite. Anomalies in this area should therefore be checked by panning the soils to see if the resistates are fresh and angular (eluvial) or smooth and rounded (alluvial).

b Peat. Large areas of upland Britain are covered in peat (a very organic-rich soil developed by anaerobic decay and compaction of vegetation) which creates an acidic reducing environment. Peat can take up metals such as lead into organic complexes, leading to high concentrations of hese elements. Levels of elements normally present in detrital minerals (such as barium in baryte) tend to be lower. Zones highly enriched in some metals may be developed close to the interface with more oxidising and alkaline environments. Determination of total organic content by ashing may be useful, though the presence of a high organic content is usually obvious when sampling. Organic-rich samples are normally avoided and samples are taken from well below the peat cover, by augering or drilling to bedrock if necessary.

c Contamination. Anthropogenic modification of near-surface soil has occurred on a large scale over the last 200 years, but is often insignificant in the more remote areas. It can cause severe problems especially in mining areas where the smelting of copper, lead, arsenic, tin and zinc has dispersed these elements over the adjacent downwind soils. Notable examples include the Snailbeach lead mine in west Shropshire, some Pennine lead mines where smelting was often carried out on an adjacent hill ('Bole hill') and south-west England. Abrahams and Thornton (1987) describe the extent and distribution of contamination in the Camborne-Hayle area of south-west England. Inefficient recovery practices have left metal-rich waste dumps which have often been levelled or used as roadstone. Floodplains may be heavily contaminated from upstream mines several miles away and, as a result, particular care should be given to the local environment of any anomalous area. The effects of lead and later zinc mining in the Alston area of the Northern Pennine Orefield on the metal contents of overbank sediments of the River Tyne and its catchments are described by Macklin and Dowsett (1989). The MRP reports are a very useful source of information on local or regional problems in the collection and interpretation of geochemical data.

Sequential leaching of soil samples, to release weakly attached metal ions from soils, can provide useful information and has been applied to the Lagalochan porphyry-style Cu-Au prospect in western Scotland using the Mobile Metal Ion (MMI) technique (Mann et al., 1998). This shows evidence of mineral zoning and provides significant geochemical anomalies in an area with high rainfall and thin peaty soils. Mobile metal ion and enzyme leach methods have not yet received rigorous testing in Britain.

G-BASE undertakes regional-scale soil sampling to supplement drainage sampling in areas of poor surface drainage. As the objective is to provide base-line data for a variety of uses a specific fit-for-purpose methodology is employed. Sites are taken on a systematic basis from alternate 1 km national grid squares with random site selection within each square. Samples comprise about 250 g of material taken from a standard depth of 30–40 cm, normally B horizon. The –150 μm fraction is extracted from the disaggregated samples and analysed for a similar wide range of elements to the stream sediments.

6.1.3 Waters

Natural waters, including surface waters and shallow groundwaters, contain information about mobile weathering products which may be used to complement stream-sediment and soil samples (Edmunds, 1971). The technique has been used for uranium, copper, fluorine and some other relatively soluble elements for many years in Britain, but the recent development of rapid, highly sensitive multi-element analytical techniques has led to new opportunities for the use of natural waters in geochemical prospecting.

Constraints include (a) the low concentrations of most elements of interest, (b) the care required to avoid contamination, (c) interpretational problems of relating anomalies to source and (d) the need to take account of terrain and hydrological conditions. Different suites of elements may be mobilised at different pH conditions and carefully designed sampling programmes can take advantage of these differences.

The G-BASE programme has regularly measured uranium, fluorine, specific electrical conductivity and pH in stream waters. More recently, multi-element and anion data for 22 variables using ICP-AES and ICP-MS have been obtained (British Geological Survey, 1999b).

6.1.4 *Soil-gas*

Soil-gas methods have been tested and used on a few mineral exploration projects in Britain, both to detect buried mineralisation and to locate fault lines. They offer simple, rapid and direct methods of determining radon, oxygen, carbon dioxide and sulphur compounds (Ball et al., 1985b). Sulphide ores tend to oxidise in aerobic conditions, consuming oxygen and releasing carbon dioxide, producing anomalous variation in the ground above seepage lines. Radon gas measurement has been used to detect uranium mineralisation and also to locate faults which may control mineral deposition (e.g. Ball et al., 1991). Response problems exist in water-logged areas where the soil gas may be affected by the decomposition of organic material and other processes.

6.1.5 *Biogeochemistry*

Various forms of biological material have been sampled for geochemical exploration purposes in Britain (e.g. Al Ajely et al., 1984) but the number of applications has been small in total. The most useful samples are probably the leaves and shoots of trees which have root systems extending for several metres into the underlying bedrock. The collection, preparation, analysis and interpretation of samples requires great care if a standardised, reproducible output is to be achieved (Smith and Ball, 1982). One of the principal problems in Britain has been identifying adequate coverage of a suitable species over the sampling area, and most applications have been in areas of recent forest monoculture.

6.2 Geophysics

6.2.1 *Regional surveys*

BGS holds nation-wide regional gravity data, which averages one station per 2 km^2. Coverage is denser in lowland Britain and sparser in highland areas, reflecting the variation in road networks and amount of Ordnance Survey (OS) levelling data. The gravity data are reduced using standard procedures to produce Bouguer gravity anomalies to an overall accuracy of 1 mGal. They provide useful baseline data for mineral exploration assessment purposes at the regional scale, providing information on major geological features including the form, depth and composition of intrusions, thickness and overall composition of sedimentary sequences, location of faults and deep fractures, and the structural framework of an area. Gravity data are used at this scale to detect features controlling the location of mineral deposits rather than the deposits themselves.

Nation-wide aeromagnetic data are also held by the BGS. These data are based on flight lines typically spaced at 2 km with a ground clearance of 305 m. The data were originally analogue and have been converted to digital form (Figure 13) using 10 nT intercepts along the flight lines (Smith and Royles, 1989). These data provide useful information on larger intrusions and other regional scale geological features that can be used in conjunction with the gravity measurements for regional structural interpretation.

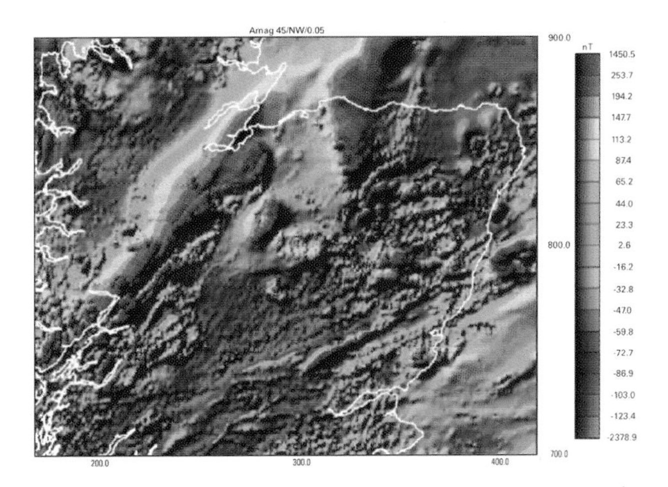

Figure 13 Regional airborne magnetic survey of central and northern Scotland.

6.2.2 *High resolution airborne surveys*

Detailed airborne geophysical surveys, with flight line spacing of a few hundred metres, ground clearance of less than 200 m, and a combination of magnetic, electromagnetic (EM) and radiometric instruments, provide high resolution data that are particularly useful for mineral exploration purposes. The BGS holds data for 26 of these surveys flown in Britain since 1957, many of them carried out as part of the Mineral Reconnaissance Programme or under the MEIGA scheme. The surveys are distributed across the country, mainly over rocks of Precambrian, Lower Palaeozoic and Carboniferous age and range in size from 6 km^2 to 14 000 km^2 (Figure 14). A summary of survey parameters are shown in Appendix 5. Significant mineral deposits within these areas include the granite-related mineralisation of south-west England, the volcanogenic massive-sulphide deposit at Parys Mountain in North Wales and strata-bound mineralisation in the Dalradian of Scotland. Many of the older analogue surveys have been converted to digital form to allow full use of interpretation software. A review of these detailed airborne geophysical surveys is given in MRP 136 whilst MRP reports 18, 20, 24, 25, 27, 29, 31, 47, 62, 65, 66 and 84 deal with specific surveys.

Two digital data sets acquired since the publication of MRP 136 are:

1 The first phase of the high-resolution resource and environmental airborne survey of the UK (HI-RES) flown over the English Midlands in 1998 (area 26 in Figure 14). This survey collected high sensitivity magnetic, VLF-EM and radiometrics data on flight lines at 400 m spacing at a height of 90 m (250 m over urban areas). A full evaluation of results has not yet been made, but it is clear that the data will be of both

regional and local importance, being able to indicate controls on mineralisation and features related to individual deposits.

2 The total count radiometric data from the 1957–9 surveys of south-west England (Areas 1 and 2 in Figure 14) have recently been digitised. Although measured with simple spectrometers, these data contain numerous anomalies related to mineralisation and old mine workings as well as artificial sources. The digital maps reveal regional features related to structure, granitic intrusions and the host rocks that are not evident on the original hand-contoured plots (Kimbell et al., 1998).

The interpretation of all airborne geophysical data in Britain may be limited locally by:

a Cultural noise. Large parts of lowland UK are built up and this can have an effect on results. EM data are particularly affected by this, for example EM data from the Anglesey survey were swamped by signals from a radio transmitter in the survey area. Research is needed at the planning stage to investigate and minimise potential problems of this nature. The BGS HI-RES project is undertaking research into different methods of removing cultural noise from airborne data (Williamson, 2000).

b Topography. Maintaining a constant terrain clearance can be difficult in highland areas and variations will affect survey results. Using a helicopter rather than fixed wing aircraft reduces this problem, but even if terrain clearance is held constant rugged topography will have an effect on geophysical data. Radiometric data in particular is affected by height variation and solid angle effects.

c Overburden. Superficial deposits affect data from airborne surveying. In the case of electromagnetic and radiometric surveys such deposits can mask any signals from bedrock. Heavily weathered bedrock can have a similar effect.

Of the airborne methods deployed, the magnetic data have proved most useful in general but in some areas such as the Dalradian of the central Highlands the EM data have proved invaluable. Apart from south-west England, the radiometric data collected during these surveys have been little used, though they have shown the capability of detecting uraniferous geological units such as dark mudstones as well as mineralisation containing uranium.

Details of data and products available from BGS are given in Section 8.1.

6.2.3 *Local ground surveys*

The BGS is developing a database of ground geophysical surveys which currently contains information on over 600 individual geophysical surveys undertaken in Britain since the early 1950's (Figure 15). Almost 500 of these surveys were carried out for mineral exploration and include follow-up to airborne surveys and geophysical investigation of geochemical anomalies, mineral occurrences and extensions of known deposits. The surveys cover a variety of environments and targets, including vein, stratabound, VMS, porphyry and stockwork mineralisation and alteration haloes. Geophysics can be used for the direct detection of mineralisation, or indirectly by defining host structures such as the buried granites in south-west England, mapped by detailed gravity surveys (MRP 34). Commercial seismic reflection data have played a major role in defining basin structures (Plant and Jones, 1999). All geophysical techniques depend on detecting variations in one or more of the physical properties of rocks (e.g. electrical properties, magnetic susceptibility and density), and appropriate exploration methods have to be selected depending on the geological target and the local environment. A number of techniques were evaluated by Bowker and Hill (1987) in an investigation of geophysical methods over the Gairloch deposit in northern Scotland. Many MRP reports contain information on and appraisal of specific ground geophysical survey methods in different geological environments in Britain (MRP 10, 16, 29, 33, 34, 35, 36, 38, 40, 56, 75, 78, 79, 85, 87, 97, 102, 112, 116, 120, 123, 127, 131, 137 and 145).

The main methods used by the Mineral Reconnaissance Programme were electromagnetic (EM), very low frequency electromagnetic (VLF-EM), induced polarisation (IP) and magnetics. EM and VLF-EM surveys were used to define geological units in a number of surveys and proved of great value in mapping Dalradian rocks, especially the resistive baryte horizons at Aberfeldy (MRP 26 and 40). IP and resistivity methods have been used widely to locate sulphides associated with a variety of types of mineralisation including VMS, stratabound and porphyry types. Supplementary information is normally required to aid interpretation of the results and locate drilling targets, as in many of the geological settings anomalies may be caused by graphite or pyrite without significant enrichment in economic minerals. In some cases magnetic anomalies can be directly related to mineralisation, for example in the manganese deposits near Rhiw in north-west Wales (MRP 102), while in others they principally aid interpretation of other data. Gravity and seismic methods have been little used at the local scale to seek individual deposits.

The application and interpretation of particular methods can be hampered in a densely populated country like Britain by the same cultural factors as affect airborne surveying. In particular, power lines, radio transmitters, wire fences, buried pipes, roads, railways and metal buildings create electrical or magnetic noise which can interfere with surveys. As well as cultural noise, ground surveys may also be influenced by:

a Overburden thickness and conductivity will have an affect on skin depths and can mask data from bedrock.

b Topography, Eg VLF-EM and radiometrics data are affected by slopes and the solid-angle effect.

Nevertheless, the MRP reports demonstrate that useful surveys can be carried out under difficult conditions, producing results, which can provide a valuable contribution to mineral exploration. Details of geophysical data and products available from BGS are given in section 8.1.6.

6.3 Remote sensing and aerial surveys

Remote sensing (air and spaceborne) techniques have been applied to various area of Britain. There is complete coverage of the country by aerial photography, some in colour, at a scale of about 1:7500, though not all scenes are cloud free. The photographs can be purchased from a number of sources listed in the UK Remote Sensing Directory which is available from the National Remote Sensing Centre (Secion 8.10.7) and also from the Ordnance Survey. They often provide useful information although analysis is complicated by large- and small-scale cultural features which can date back several thousand years, even in remote areas.

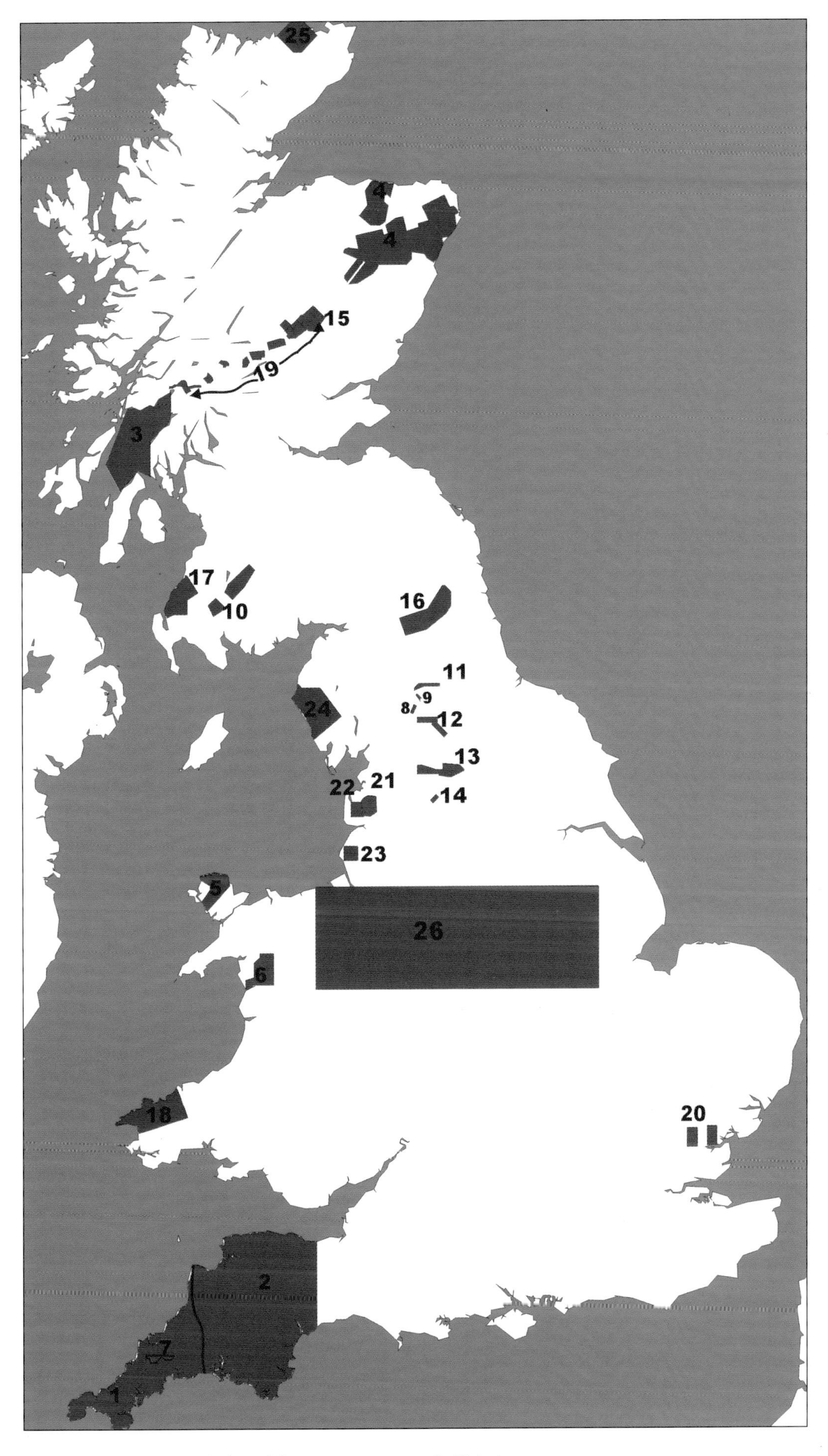

Figure 14 High-resolution airborne survey areas in Britain. Numbers refer to survey areas listed in Appendix 5.

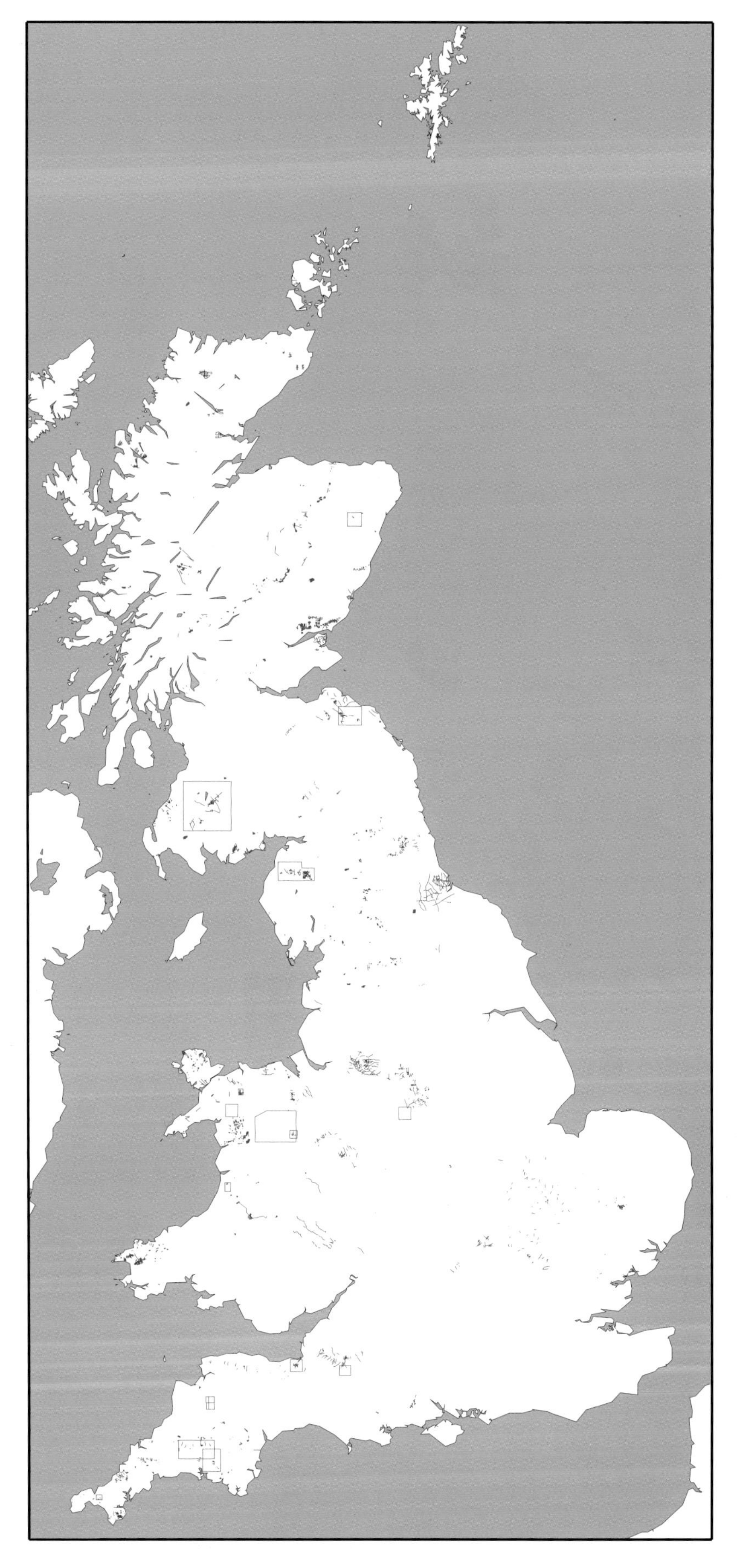

Figure 15 Ground geophysical surveys for which the BGS holds data.

Figure 16 PIMA in Parys Mountain open pit.

Satellite and airborne imagery is widely available for Britain from the National Remote Sensing Centre. There can be problems in obtaining cloud-free scenes for specific time frames, but low-sun-angle cloud-free images which are often the best for geological and structural interpretation can be obtained for most of the country. Publications on the use of remote sensing imagery in mineral exploration in Britain include Moore and Camm (1982) on structural mapping for tin-tungsten prospecting in south-west England, Hunting Geology and Geophysics (1983) on exploration for fluorite and lead-zinc in the Pennine orefields and Darch and Barber (1983) on vegetation studies over the Coed-y-Brenin porphyry copper deposit in North Wales. Greenbaum (1987) evaluated the use of airborne multispectral scanning for lithological discrimination in North Wales. In general terms the applications of satellite imagery in Britain are more limited and less effective than in semi-arid terrains. Distinction of geological units is normally difficult, with glacial deposits, cultural noise and vegetation often combining to mask geological features. Nevertheless useful geological and structural information can be obtained, particularly at the regional scale.

The PIMA (Portable Infra-red Mineral Analyser) instrument has recently been used (Figure 16) in the examination of hydrothermal alteration of the host rocks to the Parys Mountain (Colman and Cooper, 1998) and Lagalochan deposits but this has yet to be tested against satellite or airborne data.

6.4 Multi-dataset prospectivity surveys

The introduction of GIS techniques has enabled the interactive combination of all relevant geological, geochemical, geophysical and other data, followed by its interpretation in the light of ore-deposit models to produce weighted prospectivity maps for particular styles of mineralisation (e.g. Gunn et al., 1997. These methods have been applied to the characterisation of European gold deposits (Plant et al., 1998), which includes an overview of the Caledonides of northern Britain. A more detailed interpretation of the Southern Uplands of Scotland for mesothermal gold deposits has also been published (MRP 141). A study of the gold prospectivity of the Dalradian terrane is well advanced (Figure 8), and similar studies covering other parts of Britain any types of mineralisation are in progress.

7 MINERAL LEGISLATION

7.1 Mineral rights

7.1.1 England, Wales and Scotland

The rights to non-fuel minerals in Great Britain, with the exception of gold and silver, are mainly in private ownership although a significant proportion is owned by the Crown and by Government departments and agencies. Uranium and other prescribed minerals relating to the production of atomic energy belong to the mineral rights owner, but may be compulsorily purchased by the Secretary of State for Trade and Industry with compensation under powers granted in the Atomic Energy Act 1946. Although mineral rights are generally held by the surface landowner, they may have been retained by a previous landowner when the surface freehold was sold, particularly in areas with a long history of mining such as south-west England. There is no national register of mineral rights, but the Land Registry may have details of surface ownership and current ownership of mineral rights. The registers are open for public inspection. Addresses of the Land Registries of England and Wales and of Scotland are given in Section 8.9.2.

The right to exploit minerals in the foreshore (beach) and on the sea bed within the limits of national jurisdiction is vested in the Crown under the Continental Shelf Act 1964 and, apart from coal, oil and natural gas, these resources are managed by the Crown Estate Commissioners (Section 8.9.1). The only exceptions are the counties of Cornwall and Lancaster, where the foreshore is owned by the respective Duchies, and where grants of the foreshore have been made by the Crown to other parties.

The mineral rights to the noble metals, gold and silver, in most of Britain are owned by the Crown, and a licence for the exploration and development of these metals must be obtained from the Crown Estate Commissioners through the Crown Mineral Agent. There is no standard application form or licence. Applications to explore an area should be made to the Crown Estate Commissioners accompanied by a proposed work programme and details of the applicant's financial resources and technical ability. The Crown Mineral Agent then decides on the area allowed, the fee and royalty payable and the suitability of the work programme. A provisional exploration licence is usually issued for one year and can be extended. Such extention being dependent on the applicant's progress. The exploration licence can be converted into a mining lease, subject to the applicant's progress and prospects. Annual reports to the Crown Estate Commissioners are required; the data normally only remains confidential for the duration of the licence and/or lease. The exploration licence confers no rights of entry and the applicant has to negotiate access with surface rights owners and obtain planning permission from the local authority if necessary. The areas currently under licence are shown on a map (updated annually) in the United Kingdom Minerals Yearbook (Section 8.1.5). Since 1987 the results of prospecting for gold and silver have been passed to the BGS by the Crown Estate Commissioners and some are now available for public inspection along with other information from the BGS archives.

The rights to gold and silver in the former county of Sutherland in northern Scotland are held by the Duchy of Sutherland. In the Isle of Man title to all minerals, including gold and silver, is vested in the Manx Department of Industry (Section 8.7) through the Minerals Act 1986. The Department issues exploration and development licences.

The Mines (Working Facilities and Support) Act 1966, as amended, provides a means by which an operator who is either unable to trace the mineral rights owner, or cannot reach an agreement on reasonable terms with him, can obtain the necessary authority to explore for and work minerals. The Act can also be used to acquire any ancillary rights needed to facilitate the working of minerals. Although the legislation rarely has to be used, its existence is of value to prospective mineral operators as a means of persuading landowners to reach agreement. The Royal Institution of Chartered Surveyors (Section 8.9.3) has produced an informative guide to this Act (Anon, 1983a) and has also published a useful, if now somewhat dated, discussion paper on access to mineral resources in Britain (Anon, 1986).

The ownership of oil and gas in Great Britain was vested in the Crown by the Petroleum Production Act 1934. The Department of Trade and Industry grants exploration, appraisal and production licences. The rights to coal in Great Britain are vested in the Coal Authority (Section 8.10.5) by the Coal Industry Act 1994. The Authority issues licences to private operators for the working of coal from underground mines and from opencast operations.

7.1.2 Northern Ireland

All minerals in Northern Ireland, except gold and silver (already owned by the Crown) and 'common' substances, including sand and gravel and aggregates, were vested in the Department of Economic Development (now the Department of Enterprise, Trade and Investment (DETI)) by the Mineral Development Act (Northern Ireland) 1969. The Department grants prospecting and mining licences. Prospecting licences can be for up to 250 km^2 for an initial period of two years, renewable for a further two years at the discretion of the Department. Licensees are required to carry out an agreed scheme of prospecting and to report the results of their work programmes to the Department annually. This information is kept confidential for up to ten years if the company so wishes, but thereafter is available for public consultation at the Geological Survey of Northern Ireland. Exploration for gold and silver requires a licence from the DETI (Section 8.4.1) and also from the Crown Estate Commissioners, as in Great Britain. Petroleum in Northern Ireland is vested in the DETI by the Petroleum Production Act (Northern Ireland) 1964, and the Department grants licences to explore for and exploit petroleum.

7.2 Access to land

Most land is owned by the occupier or farmer who holds the surface rights. Permission must generally be obtained from the surface landowner to gain access to land for prospecting, geological mapping and geochemical and geophysical surveying. The permission of the mineral rights owner, where he is not the surface owner, is not necessary. Such an arrangement confers no rights to exploit minerals if found. Informal, even oral, arrangements may be acceptable where reconnaissance work is being undertaken, but where more costly exploration work is contemplated the company should seek a written agreement with the surface landowner. In Northern Ireland a prospecting licence from the DETI provides a right of access onto land.

The approach of companies varies, but many prefer to have a flexible legal agreement allowing surface access for prospecting, including overburden drilling and geophysics, with the right of a first refusal to an option over the mineral rights after a specified time.

The property agreement usually involves a small payment to the relevant surface and/or mineral rights owners. The establishment of local contacts or a local office is a useful step in developing an exploration programme in an area, as tracing, and negotiating with, land and mineral rights owners may sometimes be time consuming. It can be an advantage to seek out owners of major tracts such as the Crown Estate Commissioners, the Duchy of Cornwall, the Forestry Commission, sporting estates, water authorities and, increasingly, pension funds and other financial institutions. Access to quite large areas can be obtained in this way, and negotiations may be held with estate officers used to legal agreements. In Scotland the most comprehensive and publicly available source of information on estate boundaries is that compiled by Wightman (1996).

The services of a land agent experienced in mineral agreements may be advisable if large-scale exploration is envisaged. The Country Landowners Association (Section 8.9.3) has published a booklet entitled 'Minerals' (2nd Edition 1983) which provides guidelines for negotiating prospecting and mining leases from the landowner's point of view. The Institution of Mining and Metallurgy has published the proceedings of a meeting on 'The legal aspects of prospecting in the United Kingdom' (Anon, 1983b). The proceedings include a description of obtaining mineral exploration permission in south-west England and the planning application system.

The British Geological Survey has experience of working in every part of Britain. Mineral exploration companies should make early contact with the BGS to obtain background information on the geology, mineralisation and previous exploration of any area. The BGS must be informed in writing, under the Mining Industry Act 1926, of the sinking of boreholes or shafts exceeding 30 m in depth; records of the operation, including drill logs, must be kept, and the BGS must be permitted to inspect the operation and remove representative samples if it so wishes.

7.3 Planning controls on mineral operations

As in other developed countries, Britain is subject to planning controls governing most forms of 'development' of land, including mining activities, under the guidance of the Department of the Environment, Transport and the Regions (DETR). The planning framework in England and Wales is provided by the Town and Country Planning Act 1971, as amended by the Town and Country Planning (Minerals) Act 1981. Day-to-day responsibility for administering the planning system as it relates to minerals rests with the mineral planning authorities (MPAs). These are mainly the county councils, although there are exceptions in Greater London and the Metropolitan Areas. In the Peak District and Lake District National Parks, the Peak Park Joint Planning Board and the Lake District Special Planning Board, act as the MPAs.

In Scotland, the relevant Act is the Town and Country Planning Act (Scotland) 1972, as amended by the Town and Country Planning (Minerals) Act 1981. There is no separate regime for mineral planning. Proposals for mining activities are dealt with by the authority responsible for all forms of development control. In Highland, Borders, Dumfries and Galloway Regions and Orkney, Shetland and Western Isles, planning control is exercised by the Regional Council or Island Authority. Elsewhere it is the responsibility of the District Council, although Regional Councils have reserved powers related to structure planning responsibilities.

In Northern Ireland, planning permission is not usually required for normal prospecting activities. However, the Department of the Environment in Northern Ireland is responsible for planning and is informed when a prospecting licence is issued. It may require information about the proposed work.

The key feature of the planning system is that most forms of development in Britain require planning permission before development can take place. Each application is considered on its merits. Prospective developers should make their planning applications on forms provided by the planning authority. The application will be considered by a planning committee which comprises elected councillors advised by the County Planning Officer and assisted, in minerals cases, by the Minerals Officer. Developers are strongly advised to discuss their proposals with the Minerals Officer first before making any formal application. The Minerals Officer will be able to advise developers on their applications and on what supporting information will be required to help the planning committee reach a decision. The addresses of the County, Region and District planning departments, and their Minerals Officers, are listed in Harrison and Machin (1999).

In considering applications for planning permission, planning authorities will take into account the provisions of the development plan. In Great Britain, the broad framework for the use of land is provided by Structure Plans. Local mineral plans will build on this framework with more site-specific proposals. All parts of Great Britain are covered by Structure Plans; a number of Mineral Local Plans are published. Prospective mineral developers should acquaint themselves with the appropriate development plans. They set out policies for future mineral development and often contain useful information about past and current mineral working in the area. They also contain criteria against which mineral development applications will be assessed and describe policies for the restoration and after-use of mineral sites.

Although all mineral working activities come under the control of the Town and Country Planning Acts, certain operations, including most mineral prospecting activities that have little effect on the environment, do not require specific planning permission. These include drilling exploration boreholes, sinking small test pits and carrying out geophysical and geochemical surveys provided that i) the operations do not last longer than twenty eight consecutive days, ii) the work is not in environmentally sensitive areas, such as National Parks, and iii) the sites are restored soon after operations cease. The operations in this case are defined under the Town and Country Planning General Development Order 1988 (GDO) and planning permission is assumed to have been granted.

If the developer has notified the planning authority in advance then the twenty eight day period is extended to four months. Mineral developers should discuss their intended exploration activities with the relevant Minerals Officer to determine if their proposals fall within the scope of the GDO.

After the planning committee has considered the planning application they may decide to approve it, refuse it, or approve it subject to certain conditions. In practice most permissions for mineral development have conditions attached which are designed to control their impact on the environment. For example, conditions may be imposed which control the hours of working and access arrangements, set noise limits and limit the depth of working. Other conditions may govern the restoration of the site.

If the application is refused, or the conditions are unacceptable, the applicant has the right of appeal to the Secretary of State for the Environment, Transport and the Regions. An appeal following refusal of a minerals application usually results in a public inquiry, conducted by a planning inspector, in order that all individuals or organisations with a genuine interest may be given the opportunity of presenting their case. The inspector has delegated powers to determine the appeal. However, in certain circumstances, the Secretary of State will make the decision, taking account of the inspector's recommendations.

The Secretary of State also has the right to 'call in' planning applications. In practice, it is established policy that minerals cases will be called in only if they raise issues of regional or national importance. Before determining a called-in case, the Secretary of State will hold a public inquiry in a manner similar to an appeal case. One recent called-in application was that for the Hemerdon tungsten-tin open pit deposit on the border of the Dartmoor National Park. The application was approved, subject to certain conditions. Others, which have been granted full planning permission, include the Cononish and Cavanacaw gold mines. The major Duntanlich underground baryte mine, near Aberfeldy, was refused permission in 1997 and again on appeal in 1998.

Applications for mineral working in some environmentally sensitive areas such as National Parks (NP), Areas of Outstanding Natural Beauty (AONB) and Sites of Special Scientific Interest (SSSI) will be subject to the most rigorous examination. However, large-scale mineral operations for china clay, fluorspar, limestone, potash, roadstone and slate are carried out within National Parks from open-pit and underground mines on a basis of national need, indigenous industry and local employment.

A considerable amount of mineral exploration has been carried out in Britain in recent years by numerous British-based and overseas companies. The domestic minerals industry and the planning authorities have built up a co-operative atmosphere over the years. However, companies without any experience of operations in Britain would be well advised to seek professional advice before attempting to carry out work likely to require planning permission.

The DETR has commissioned the BGS to produce a series of county-based maps and explanatory texts under the general heading 'Mineral Resource Information for Development Plans'. These show the extent of constructional, coal and selected industrial mineral resources on a topographic background together with existing mineral planning permissions for mineral extraction and constraint areas, such as National Parks and Sites of Special Scientific Interest (SSSI). These are produced as paper maps at a scale of 1:100 000, but the information will be available digitally through the MINGOL GIS system (see Section 8.1.8). Figure 17 contains an extract from one of the maps, showing mineral resource areas, mineral planning permissions, active mines and quarries, and constraint areas.

Further useful information on minerals planning can be found in the Minerals Planning Guidance Notes (MPGs) issued by the Department of the Environment from 1988. These set out Government policy and provide updated information on the 1981 Minerals Act.

8 SOURCES OF INFORMATION

8.1 British Geological Survey

The major source of information for those engaged in mineral exploration is the British Geological Survey (BGS).

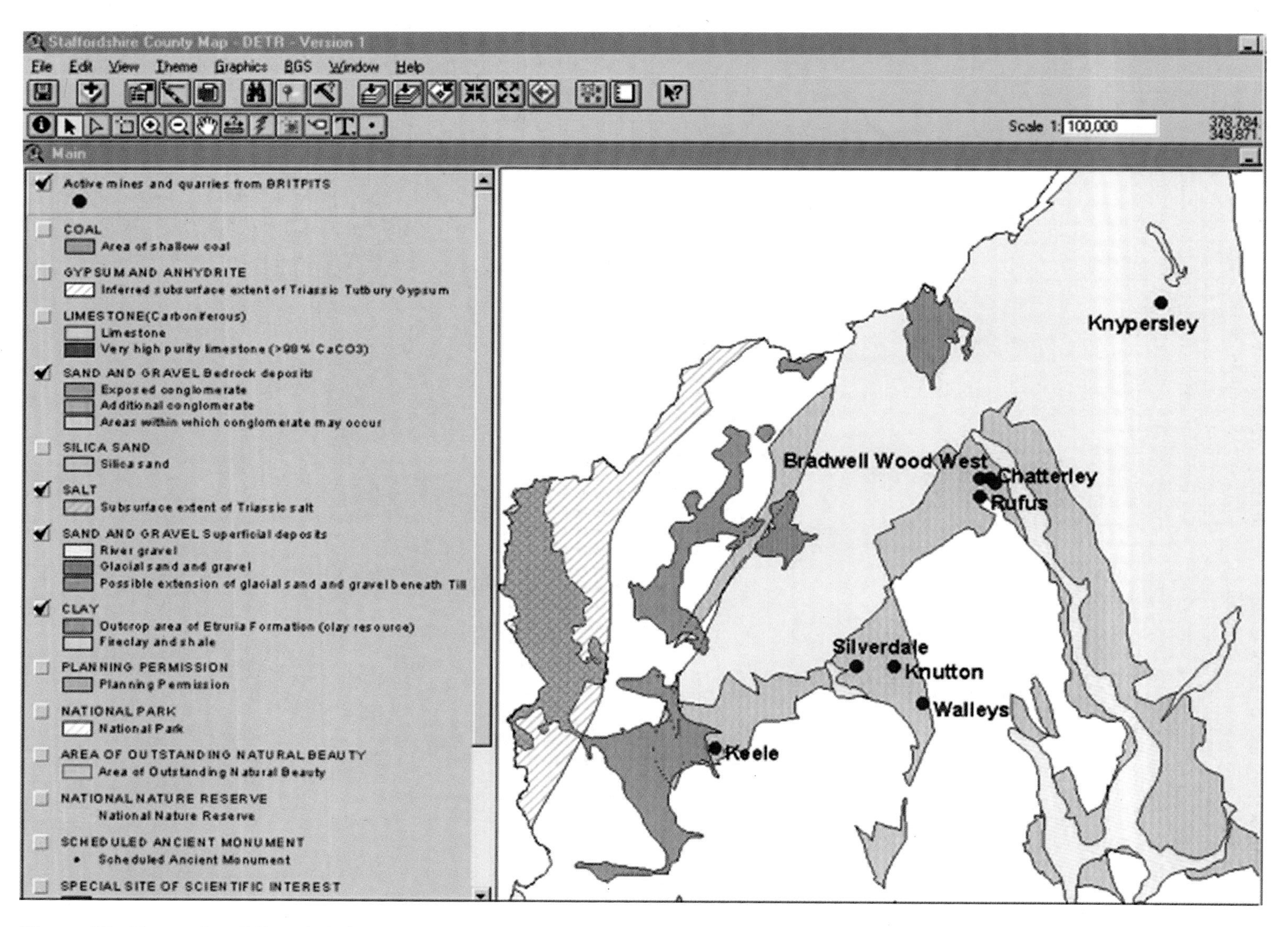

Figure 17 Extract from Minerals Information for Development plan for Staffordshire in GIS form.

The BGS is the main Government agency in Great Britain for undertaking national work in the earth sciences and is the recognised national repository for geoscience data (see section 2). It produces geological maps based on primary surveys normally carried out at 1:10 000 scale. These are compiled into published 1:50 000 geological maps of both solid (hardrock) and superficial (drift) formations. The BGS publishes a thematic map series at 1:1 m or 1:1.5 m scales. These include a metallogenic map of Britain and Ireland (British Geological Survey, 1996a), tectonic (British Geological Survey, 1996c) and geophysical (gravity and aeromagnetic) maps of Britain, Ireland and adjacent areas, an industrial minerals map of Britain (British Geological Survey, 1996b) showing the distribution of extraction sites and, most recently, a coal resources map of Britain (British Geological Survey, 1999a). Geophysical, geochemical and thematic map series are also produced and much of the data on the databases behind these maps can be purchased in digital form. The BGS holds very large amounts of geological information including samples, borehole cores, mining and mineral exploration records, plans and old maps. It maintains a central library which has over 500 000 volumes and pamphlets and 180 000 maps, and receives over 3000 journals. It carries out specialist functions for Government departments and private companies, such as geochemical and geophysical investigations, as well as mineral resource evaluation and environmental geology studies. A catalogue of published maps, memoirs and other publications is available. The BGS work and products of most interest to those undertaking mineral exploration are outlined below:

8.1.1 Geological maps, memoirs and special reports

Detailed geological mapping has been a primary function of the BGS since its foundation in 1835 and there is a constant programme of revision and remapping. Printed maps are available at 1:625 000, 1:250 000 and 1:50 000 with some at 1:25 000. Maps at larger scale (1:10 000 or 6 inches to the mile) can usually be obtained but some may be copies of old, unpublished manuscript maps, which date from the last century. The geological maps of Great Britain all use an Ordnance Survey topogrpahic base and the most recent maps are held in digital form. Descriptive memoirs exist for many of the 1:50 000 maps and these can be a fruitful source of information on mineral occurrences. For example, the Gairloch discovery outcrop was noted in the 1907 Geological Survey Memoir of the area as 'a copper-bearing limestone' (Peach et al., 1907). Follow-up of this reference in 1978 led directly to the identification of the Gairloch Cu-Zn-Au deposit (Jones et al., 1987).

Economic memoirs and related publications provide minerals information on some of the old mining fields of Britain, notably the North Pennines, west Cumbria and south-west England. A series of Special Reports on the Mineral Resources of Great Britain were published by the Geological Survey from 1915 to 1925. They cover most of the metallic and related minerals and, although dated, do contain much useful information, as many more mines were working at this time. There is also a series of mineral dossiers produced in the 1970s that provide data on a number of metallic and industrial minerals including baryte, fluorspar, tungsten, tin, gold and silver. Many of these reports are listed in the references.

8.1.2 Mineral Reconnaissance Programme (MRP) data and reports

The MRP was funded by the Department of Trade and Industry between 1973 and 1997 to provide baseline geological, geochemical, geophysical and metallogenic information on potentially prospective areas in Great Britain that would encourage private-sector investment. The work was at various levels, from initial reconnaissance to diamond drilling, but did not go beyond the discovery stage of a mineral deposit. Although work was generally aimed at specific styles of mineralisation, a wide range of elements was determined and a variety of techniques used. The results are published in a series of over 150 reports and data releases. A summary of the first 109 reports is given by Haslam et al. (1990) who also list many other useful papers. The areas covered by the MRP are shown in Figure 18, and Appendix 2 gives the report titles

8.1.3 BGS Minerals Programme

The MRP was subsumed into a new BGS Minerals Programme in 1997. This programme contains two key elements:

- Minerals Information and Advisory Service. This draws heavily on DTI-supported minerals databases, to provide information and advice to government and industry on minerals extraction and usage, and related issues. The information assists in the determination of government and corporate policy and mineral resource management, including decisions related to the balance between the need to develop natural resources and the protection of the environment.
- Promotion and enhancement of mineral wealth. This is aimed at stimulating private sector investment in the optimal development and use of Britain's mineral resources, and the transfer to industry of related enabling technology (especially IT). This is achieved by the production of best-practice manuals, prospectivity maps and reports and other information that will reduce entry costs for SMEs exploring for minerals in Britain.

The Programme maintains extensive databases that include information on world mineral production, availability and trade, and exploration data collected by the MRP. The Programme publishes 'World Mineral Statistics' annually and reports on mineral exploration prospects and methods in Britain. A new series of brochures 'Minerals in Britain' is also being produced with those covering gold, gemstones, lead and zinc and copper now available. The Minerals Programme website www.mineralsuk.com contains current information on mineral exploration in Britain.

8.1.4 Mining company exploration data

Mining companies are not required to deposit the results of exploration in Great Britain with a Government agency, so records of prospecting work are very incomplete. Some landowners may have the results of past exploration programmes but there is no central register for this information. However, from 1972 to 1984 the Department of Trade and Industry gave grants under the Mineral Exploration and Investment Grants Act 1972 (MEIGA) for mineral exploration for the ores of non-ferrous metals, fluorspar, barium minerals and potash, provided the results were deposited with the BGS. The results of over 150 projects submitted under the Act have now been put on Open File and are available for public inspection at the National Geoscience Records Centre (see below) and relevant regional offices of the BGS. Figure 20 and Appendix 3 give the location of

Figure 18 MRP report areas.

Open File areas, project names and metals sought. In addition, from 1999 a number of completed precious metal exploration projects, carried out under Crown Estate Commissioners Mines Royal licence, will also be placed on Open File at the National Geoscience Records Centre. Additional projects will follow in due course.

In Northern Ireland, all mining company exploration data are deposited annually with the Geological Survey of Northern Ireland. All such data remain confidential, if requested, for a period of five years, renewable on application for a further five years. Details of all non-confidential data can be obtained from the GSNI.

8.1.5 *Geochemical Baseline Survey of the Environment (G-BASE) data and atlases*

The G-BASE programme, formerly the Geochemical Survey Programme (GSP), is funded by the Office of Science and Technology to undertake the systematic geochemical mapping of Great Britain. The work started in northern Scotland and has progressed southwards as far as the English Midlands. The programme employs stream-sediment sampling at a density of one sample per km^2 with analysis for up to 30 elements. Water, soil and panned stream-sediment samples are also collected. Data accompanied by an explanatory text are presented in atlases and are also available as digital files which can be requested in various formats, e.g. by area, element or groups of elements or specified ranges for an element. Interactive, multicomponent digital image processing of the datasets enables the geochemical characteristics of specific formations to be quickly established and searches carried out for geochemical patterns associated with various styles of mineralisation. The areas covered are shown in Figure 12.

8.1.6 *Geophysical data*

Regional Data

The BGS possesses regional-scale gravity and aeromagnetic coverage of the complete land area of Britain. The data are published at the scale of 1:250 000, 1:625 000, 1:1 M and the new 1:1.5 M colour-shaded relief maps. Digital data are sold on request, either in raw form, as grids, or as maps or images of specific areas processed to the customer's requirements. Geophysical Information maps (GIM) are produced at a scale of 1:50 000, compatible with the geological map series, and show contours of gravity and magnetic anomaly (with sites of observation), locations of boreholes, and sites of detailed ground geophysical investigation.

High Resolution airborne data

A number of areas of the UK have been surveyed in detail with airborne methods as part of the MRP or other projects, including various combinations of magnetic, EM, VLF and radiometric sensors (Figure 14). Much of these data were originally analogue but have been converted to digital form in recent years. A review of detailed airborne surveys and data availability is given in MRP 136. The two latest digital datasets acquired are the recently digitised total count radiometric data from south-west England and the high resolution airborne resource and environmental survey of the UK (HI-RES) flown over the Midlands in 1998 (Figure 14). Appendix 5 contains summary information for detailed airborne survey datasets held by the BGS.

Ground Surveys and other data

There is a wealth of ground geophysical survey information collected for the MRP and many other projects. Most were collected in analogue form, but the BGS has a digital index giving details of individual surveys and lines (Figure 15). Other relevant information includes borehole geophysical logs, physical properties database, heat flow determinations and seismic reflection data.

8.1.7 *National Geosciences Data Centre*

The National Geoscience Data Centre (NGDC) is based at Keyworth and Edinburgh and contains a wide variety of information including borehole cores, cuttings, records, fossils, rock samples, thin sections and paper records. All this material is available for public inspection, subject to confidentiality restrictions. The borehole index is now available on CD-ROM. Regional offices of the BGS at Edinburgh in Scotland and Exeter in south-west England contain additional material relating to their area. A new office may be opened in Cardiff.

8.1.8 *Other minerals-related products*

Many unpublished reports on mineral exploration carried out by the BGS for clients more than 20 years ago, especially for uranium, are held on open file and are listed in Institute of Geological Sciences (1984).

Minerals-related information held by BGS is being systematically digitally indexed or converted to digital form in its entirety and linked using a GIS called MINGOL (Minerals GIS on-line). The system (Figure 19) currently holds information on active mines and quarries, mineral exploration areas, planning constraints and mineral occurrence data for parts of Britain and can be used to produce output in a variety of formats, such as reports, maps and tables, to suit customer requirements. New information is being added on a continuous basis.

The BGS Directory of Mines and Quarries, available as a book or on CD-ROM, provides up-to-date information on active mines and quarries in Britain. The data held includes location, material extracted, operator and geology.

The United Kingdom Minerals Yearbook has been produced annually by the BGS with the assistance of funding from the DETR and DTI. It provides production, consump-

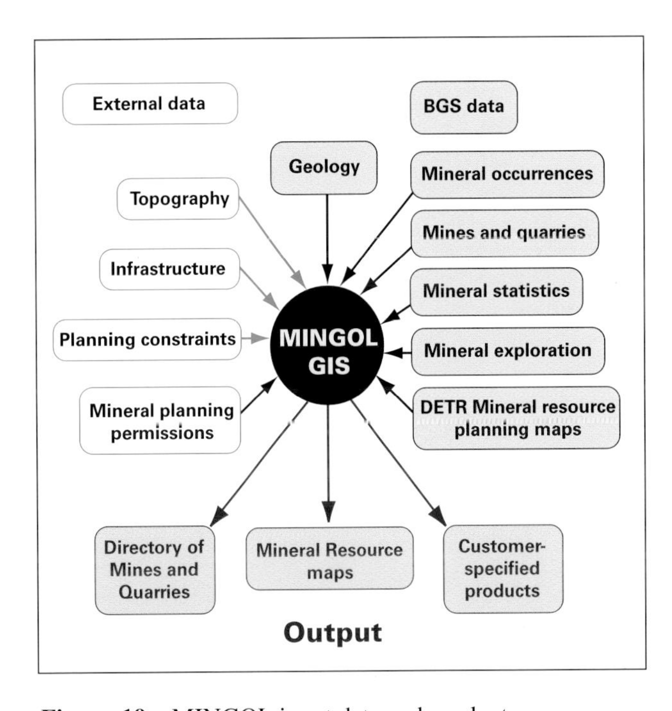

Figure 19 MINGOL input data and products.

Figure 20 MEIGA report areas

tion and trade statistics for minerals in Britain as well as a commentary on events during the year that includes a section on metalliferous mining and exploration. A map of gold licence areas and a list of companies holding licences is included. The United Kingdom Minerals Industry (Harris, 1995) gives a useful account of the constructional and industrial mineral industry in Britain with some information on metalliferous minerals.

8.2 Department of Trade and Industry (DTI)

The DTI is the interface between Government and the metalliferous and industrial minerals mining industry. It publishes "A Guide for Business" summarising its work and giving a comprehensive range of addresses and telephone numbers of contacts. Among the contacts listed are those concerned with non-energy, non-construction materials; overseas investment in Britain ("inward investment"); industry in the English regions, Scotland and Wales; consultancy assistance for smaller firms; business statistics (though the BGS is a more specialised source of mineral statistics); company information; research and technology; and selective financial assistance for firms in particular parts of the country, known as "Assisted Areas".

Department of Trade and Industry
Metals, Minerals and Shipbuilding Directorate
151 Buckingham Palace Road
London SW1W 9SS
Tel: 0207 215 1102
Fax: 0207 215 1070
Website: www.dti.gov.uk

8.2.1 The Invest in Britain Bureau (IBB)

The IBB, a joint Department of Trade and Industry and Foreign Office body, is the main Government agency for inward investment and promotes the whole of the UK as an investment location. The IBB can assist firms with all aspects of locating or relocating a business in the UK or expanding existing facilities. It is the central contact point in Britain for all advice and assistance. The Bureau operates overseas through British Embassies, High Commissions and Consulates-General. It can provide:

- information to enable firms to choose the best possible location for an operation taking account of all their requirements: availability of labour, transport, proximity to suppliers and, above all, particular commercial needs.
- information about how to set up an operation in Britain.
- information and help on the national, regional and local incentives available to encourage investment.
- help in any contact with public authorities, whether central government, local government, nationalised industries or essential services such as gas, electricity, water and telecommunications.
- visits to sites and buildings available in any part of Britain.
- contacts with key private and public sector companies.
- liasion with Regional Investment Agencies to provide a seamless delivery service to prospective inward investors.

Invest in Britain Bureau
Department of Trade and Industry
1 Victoria Street
London SW1H 0ET
Tel: 0207 215 2501
Fax: 0207 215 5651
E-mail: invest.britain@ibb.dti.gov.uk
Website: www.dti.gov.uk/ibb

8.2.2 Regional Offices

Government Office for London
Riverwalk House
157-161 Millbank
London SW1P 4RR
Tel: 0207 217 3222

Government Office for the North-east
Wellbar House
Gallowgate
Newcastle upon Tyne NE1 4TD
Tel: 0191 202 3840

Government Office for the North-west
Sunley Tower
Piccadilly Plaza
Manchester M1 4BA
Tel: 0161 952 4000

Government Office for Yorkshire and the Humber
25 Queen Street
Leeds LS1 2TW
Tel: 0113 280 0600

Government Office for the East Midlands
The Belgrave Centre
Stanley Place
Talbot Street
Nottingham NG1 5GG
Tel: 0115 971 9971

Government Office for the West Midlands
77 Paradise Circus
Queensway
Birmingham B1 2DT
Tel: 0121 212 5050

Government Office for the Eastern Region
Building A
Westbrook Centre
Milton Road
Cambridge CB4 1YG
Tel: 01223 346700

Government Office for the South East
Bridge House
1 Walnut Tree Close
Guildford
Surrey GU1 4GA
Tel: 01483 882255

Government Office for the South West
The Pithay
Bristol BS1 2PB
Tel: 0117 900 1700

Northern Ireland Region
Department of Enterprise, Trade and Investment
Netherleigh
Massey Avenue
Belfast BT4 2JP
Tel: 0232 529900

Welsh Office
Industry and Training Department
Cathays Park Cardiff CF1 3NQ
Tel: 02920 825111

Scottish Office
Meridan Court
5 Cadogan Street
Glasgow G2 6AT
Tel: 0141 248 2855

8.3 Department of the Environment, Transport and the Regions (DETR)

The DETR is the interface between Government and the bulk and construction materials mining industry. It is also responsible for the development and implementation of the planning regulations. It places contracts with the BGS and private-sector companies to provide minerals-related information as an aid to decision making at national, regional and local levels. Currently it supports the production of the United Kingdom Minerals Yearbook by the BGS. A series of maps of the counties of England and Wales at 1:100 000 scale, with accompanying reports, relating mineral resources to planning constraints is also being produced. The information from these maps is being transferred to the MINGOL GIS and will be available for producing customised local and regional maps for a variety of purposes.

Department of the Environment, Transport and the Regions
Minerals and Land Reclamation Division
Eland House, Bressenden Place
London SW1E 5DU
Tel: 0207-890 3000
E-mail: minerals_waste@detr.gov.uk
Website: Website www.detr.gov.uk

8.4 Northern Ireland

On Thursday 2 December 1999 power was devolved to the Northern Ireland Assembly and its Executive Committee of Ministers.

Northern Ireland Assembly
Castle Buildings
Stormont Estate
Belfast BT4 3SR
Tel: 02890 528400
Website: www.northernireland.gov.uk

8.4.1 Department of Enterprise, Trade and Investment

In Northern Ireland mineral exploration and development is carried out under licence from the Department of Enterprise, Trade and Investment (DETI). The Minerals and Petroleum Unit of the DETI publishes a triennial report, with annual updates, on Mineral Exploration and Development in Northern Ireland. The 1994–1997 edition also provides details of previous work carried out since the passing of the Mineral Development Act (Northern Ireland) 1969. The report includes a useful summary of the geology of Northern Ireland, the Department's policies and the addresses of companies holding exploration licences. The Geological Survey of Northern Ireland publishes a mineral localities map at a scale of 1:253 440 and maintains an associated mineral locality card index. Details of mineral localities in the Dalradian rocks of Northern Ireland and contiguous areas in the Republic of Ireland have been published (Legg et al., 1985). The Minerals and Petroleum Unit will provide general information, engage in discussion about licencing and receive expressions of interest. Technical information and advice, including non-confidential information from previous exploration projects, may be obtained from the Geological Survey of Northern Ireland. The contact details for the GSNI are given in the Summary.

Department of Enterprise, Trade and Investment
Energy, Minerals and Petroleum Unit
Netherleigh
Massey Avenue
Belfast BT4 2JP
Tel: 02890 529900
Fax: 02890 529550
Website: www.nics.gov.uk/eti.htm

8.5 Scotland

Following the recent devolution legislation, Scotland is becoming increasingly autonomous, particularly in all matters likely to be of interest to mineral exploration companies (e.g. industry, planning and development). Administration is run from the Scottish Office, which controls a number of departments and agencies.

Scottish Office
St Andrews House
Regent Road
Edinburgh EH1 3DG
Tel: 0131 556 8400
Website: www.scotland.gov.uk

8.6 Wales

Similarly, the Welsh Office has responsibility for many statutory functions in Wales, including town and country planning, land use, water, and financial assistance to industry.

Welsh Office
Crown Buildings
Cathys Park
Cardiff CF1 3NQ
Tel: 02920 825111
Website: www.wales.gov.uk

8.6.1 Welsh Development Agency

Established by the UK government in 1976, the Welsh Development Agency (WDA) has been tasked with regenerating the economic prosperity of Wales in the wake of heavy industry decline. The WDA is committed to helping existing businesses expand and thrive and to encouraging inward investment, the relocation of industry and the creation of jobs. Its success has transformed Wales into the best business climate in Europe and positioned the country at the leading edge of innovation and technology.

Client Services
WDA
Main Avenue
Treforest
Mid Glamorgan CF10 3FE, UK
Tel: 01443 845500
Fax: 01443 845589
Website: www.wda.co.uk

8.7 The Isle of Man

The Isle of Man has its own legislative assembly and is not part of Britain but has a special relationship with it.

However, it is geologically similar to adjacent areas of Britain (north-west England) and a number of Pb-Zn vein deposits have been worked in the Lower Palaeozoic rocks which underlie most of the island. Several Caledonian granite intrusions also outcrop. The economic geology of the island, including descriptions of the major mines and numerous occurrences, is described by Lamplugh (1903) who compiled his memoir at a time when the mines were still active. The two largest mines were Foxdale and Laxey which produced a total of 200 000 t of lead and 125 000 t of zinc — the latter metal almost entirely from Laxey (Webb, 1980). The lead ore was notably silver-rich with recorded contents varying from 9 oz to 70 oz (270 to 2100 g) per ton of 'lead ore' (Lamplugh, 1903). The famous Laxey waterwheel (now a tourist attraction) was used to dewater the Laxey mine. The island is thus as prospective for mineralisation as parts of the Lake District and is included in this guide for completeness. The Isle of Man Government licences and controls all mineral exploration and exploitation in the island under the Minerals Act 1986. Licences are issued by the Department of Trade and Industry. Geological maps of the island and adjacent sea bed are published by the BGS.

Department of Trade and Industry
Illiam Dhone House
2 Circular Road, Douglas IM1 1PJ, Isle of Man
Tel: 01642 685675
Fax: 01642 685683
E-mail: dti@gov.im
Website: www.gov.im

8.8 The Channel Islands

The Channel Islands are also not part of the United Kingdom but have a special relationship with it. Guernsey and Jersey have their own legislative assemblies but, like the Isle of Man, rely on Britain for some functions. Geological maps of the islands and adjacent sea bed are published by the BGS. A small silver-lead vein in Late Precambrian granite was worked on the island of Sark in the 1830s and 1840s (Ixer and Stanley, 1983).

Website: www.jersey.gov.uk and www.guernsey.gov.uk

8.9 Other sources of information

8.9.1 Mineral rights

Gold and silver exploration and mining licences are dealt with on behalf of the Crown Estate Commissioners by:

The Crown Mineral Agent
Wardell Armstrong
Lancaster Building
High Street
Newcastle-under-Lyme
Staffordshire ST5 1PQ
Tel: 0178 261 2626
Fax: 0178 266 2882
E-mail: wardell@wardell-armstrong.com

The Crown Estate Commissioners also hold the mineral rights to substantial areas in parts of England and Wales as does the Duchy of Cornwall, especially in south-west England. The relevant contacts are:

The Agricultural Estates Manager
Crown Estate Commissioners
16 Carlton House Terrace
London SW1Y 5AH
Tel: 0207 210 4377
Fax: 0207 210 4236
Website: www.crownestate.co.uk

Ms L Bryant
Property Services Manager
Duchy of Cornwall Office
The Old Rectory
Newton St Loe
Bath BA12 9BU
Tel: 0122 587 4194
Fax: 0122 587 4171

The Forestry Commission, which is the national forestry authority, operates throughout England, Scotland and Wales. Information regarding mineral development on Forestry Commission land can be obtained from:

Forestry Civil Engineering
Forestry Commission
Greenside
Peebles EH45 8JE
Tel: 0172 1720 218
Website: www.forestry.gov.uk

Information on land ownership in Scotland is provided by Wightman (1996). The larger landowners are listed and in most cases these will also be mineral rights holders.

8.9.2 Land Registry

Information on land ownership and registration services are provided by:

England and Wales:

HM Land Registry
32 Lincoln's Inn Fields
London WC2 3PH
Tel: 0207 917 8888
Fax: 0207 995 0110
Website: www.open.gov.uk/landreg

Scotland:

Registers of Scotland
Meadowbank House
153 London Road
Edinburgh EH8 7AU
Tel: 0131 659 6111
Fax: 0131 479 1221
Website: www.rosdir.gov.uk

Northern Ireland:

Land Registers of Northern Ireland
Lincoln Building
27–45 Great Victoria Street
Belfast BT2 7SL
Tel: 02890 251515/6
Fax: 02890 251 655
Website: www.doeni.gov.uk/land.htm

8.9.3 *Institutions with interests in mineral exploration*

The Institution of Mining and Metallurgy (IMM)

The IMM is the professional body representing mining and exploration geologists, engineers and metallurgists in Britain and overseas. It publishes quarterly transactions in each of the three disciplines: A, Mining industry; B, Applied earth sciences; and C, Mineral processing and extractive metallurgy. It holds numerous conferences throughout the world and publishes the proceedings.

Institution of Mining and Metallurgy
Danum House
South Parade
Doncaster DN1 2DY
Tel: 01302 380900
Fax: 01302 340554
E-mail: hq@imm.org.uk
Website: www.imm.org.uk

A London information office and the library are maintained at:

Institution of Mining and Metallurgy
77 Hallam Street
London W1N 5LR
Tel: 0207 580 3802
Fax: 0207 436 5388
E-mail: london@imm.org.uk

Mineral Industry Research Organisation (MIRO)

MIRO is a research and information trade organisation for the Minerals Industry. Its wide corporate membership geographically covers not only Britain and Europe but also North America and Africa. The members interests range from exploration through mining and mineral processing to extractive metallurgy. Its main functions are to provide the means for co-operation amongst its members to solve technical problems they have in common, to provide information and data for the industry, and to provide an industry-based focus for research and comment. It has assisted some conceptual mineral exploration programmes by organising industrial consortia to provide funding, often in conjunction with national Government or international agencies. The MIDAS pan-European gold exploration project (Plant et al., 1998) was assisted by MIRO.

Mineral Industry Research Organisation
Expert House
Sandford Street
Lichfield
Staffordshire WS13 6QA
Tel: 01543 262 957
Fax: 01543 262 183
E-mail: miro@mirolich.demon.co.uk
Website: www.miro.co.uk

The Mining Association of the United Kingdom (MAUK)

This association is subscribed to by companies et al. with interests in mining and exploration. The Association assists members with the interpretation of new legislation, both national and international, and acts as a focus for industry's interaction with Government. The Association has produced a report on the British mining industry entitled: 'Mining in the UK — The Industry and its Competitiveness'. The report gives an overview of the whole mining industry, including producers and products, and its profitability. It also examines the problems currently facing the industry and their possible solution. The report is available from the secretary of the Mining Association at the Mineral Research Organisation (MIRO) whose address is given above.

The Geological Society

The Geological Society of London is the professional body representing the interests of geologists in Britain. It publishes the Geologists Directory — see under publications above.

The Geological Society
Burlington House, Piccadilly
London W1V 0JU, UK.
Tel: 0207 434 9944
Fax: 0207 439 8975
E-mail: enquiries@geolsoc.org.uk
Website: www.geolsoc.org.uk

Mineral Planning

The quarterly publication Mineral Planning provides information on all minerals-related planning inforamtion, including applications, decisions and appeals, as well as a wide range of articles on specific topics. It also publishes the listing of mineral and waste planning officers and authorities in Great Britain (Harrison and Machin, 1999).

Mineral Planning
2 The Greenways
Little Fencote
Northallerton DL7 0TS
Tel/Fax: 0160 974 8709

The Royal Institution of Chartered Surveyors (RICS)

RICS can provide assistance in the acquisition of surface and mineral rights and in negotiations with planning authorities. The Institution publishes a booklet 'Directory of Minerals Surveying Services' (2nd Edition, 1988).

Royal Institution of Chartered Surveyors
12 Great George Street
Parliament Square
London SW1P 3AD
Tel: 0207 222 7000
Website: www.rics.org.uk

Country Landowners Association

The booklet 'Minerals' (2nd Edition, 1983) published by the Country Landowners Association gives information on the legal aspects of negotiating mineral prospecting and development agreements from the landowner's point of view.

Country Landowners Association
16 Belgrave Square
London SW1X 8PQ
Tel: 0207 235 0511
Fax: 0207 235 4696
E-mail: mail@cla.org.uk
Website: www.cla.org.uk

8.9.4 *Environmental and conservation bodies*

Environment Agency

Formed in 1996 and replacing the National Rivers Authority and pollution and waste authorities, this organisation is responsible for the enforcement of pollution control legislation, water resources mangagement, waste management policy, fisheries, flood defences and navigation.

Environment Agency
Rio House
Aztec West
Almondsbury
Bristol BS12 4UD
Tel: 01454 624400
Fax: 01454 624409
Website: www.environment-agency.gov.uk

English Nature

English Nature is the Government-funded body whose purpose is to promote the conservation of England's wildlife and natural features. This is achieved by the organisation taking actions itself and by working through and enabling others. English Nature is governed by a Council appointed by the Secretary of State for the Environment:

National Office
English Nature
Northminster House
Peterborough PE1 1UA
Tel: 01733 455000
Fax: 01733 568834
Website: www.english-nature.org.uk

The Countryside Agency

The Countryside Agency is a new statutory body resulting from the merger of the Countryside Commission and the Rural Development Commission. The Agency's main tasks are to conserve and enhance the countryside, to promote social equity and economic opportunity for the people who live there and to help everyone, wherever they live, to enjoy this national asset. The Agency is also responsible for designating specially protected areas, such as National Parks and Areas of Outstanding Natural Beauty, and advises Government on policy towards them.

The Countryside Agency
John Dower House
Crescent Place
Cheltenham
Gloucestershire GL50 3RA
Tel: 01242 521381
Fax: 01242 584270
Website: www.countryside.gov.uk

Scottish Natural Heritage

Responsible for National Scenic areas and National Nature Reserves, this organisation advises government et al., commissions and conducts relevant research and aims to promote enjoyment and understanding of the environment. It is an independent organisation but is wholly funded by government.

Scottish Natural Heritage
12 Hope Terrace
Edinburgh EH9 2AS
Tel: 0131 447 4784
Fax: 0131 446 2277
Website: www.snh.org.uk

Countryside Council for Wales

This is the government's statutory advisor on wildlife, countryside and maritime protection matters in Wales.

Countryside Council for Wales / Cyngor Cefn Gwlad Cymru
Plas Penrhos
Penrhos Road
Bangor
Gwynedd LL57 2LQ
Tel: 01248 385500
Fax: 01248 355782
Website: www.ccw.gov.uk

Environment and Heritage Service (Northern Ireland)

This is an agency within the Department of the Environment of Northern Ireland, which aims to protect and conserve the natural and built environment and to promote its appreciation for the benefit of present and future generations. Its duties include managing nature reserves, designating areas of outstanding natural beauty and special scientific interest, maintaining country parks, monitoring water quality, and controlling effluent discharges and air pollution.

Environment and Heritage Service
Commonwealth House
35 Castle Street
Belfast BT1 1GU
Tel: 02890 546533
Fax: 02890 546660
Website: www.ehsni.gov.uk

8.9.5 Universities

Many British universities and other institutions carry out basic and applied research on topics related to mineralisation. A listing of these topics, by institution and discipline, can be found in Current Research in Britain which is available from:

The British Library Document Supply Centre
Boston Spa
Wetherby
West Yorkshire LS23 7BQ
Tel: 01937 546000
Fax: 01937 546333
Website: www.bl.uk

More general information on British Universities and their staff can found in the Commonwealth Universities Yearbook and the publications of individual universities.

8.10 Publications

8.10.1 General

There are a number of general sources of information on mineral deposits and mineral exploration in Britain. Dunham et al. (1979) provided a comprehensive account of the known mineralisation of Britain in 'Mineral Deposits of Europe Volume 1: North-west Europe', part of the Mineral Deposits of Europe series. 'Mineralization in the British Isles', edited by Patrick and Polya (1993), is an important compilation of descriptions of the metallogeny of the principal mining areas in Britain and Ireland. Individual chapters are referred to throughout this publication. Webb (1980) published a mineral deposit map of Great Britain on two sheets at 10 miles to 1 inch scale. It included the general location of all the major and most of the minor mineral deposits together with production statistics and a comprehensive bibliography to 1980. A recent comprehensive account of the minerals of Wales, with a summary mining history and extensive bibliography is given by Bevins (1994). Beveridge et al. (1991) provide a review of metalliferous mining in Scotland. A summary of metal mining activities in Britain in the 1970s is given in Anon. (1982). Tylecote (1986) has written a study of the history of metallogeny in Britain, with information on gold, copper, tin, lead, silver and iron production since earliest times to the medieval period. Richardson (1974) provides a historical overview of metalliferous mineral development in Britain from then to the present.

A number of journals specialise in papers relating to the history and archaeology of mining in Britain which may be relevant to mineral exploration. They include British Mining, published by the Northern Mines Research Society, the Bulletin of the Peak District Mines Historical Society, the Journal of the Russell Society and the UK Handbook of Mines and Minerals. These various journals, and the societies which publish them, can most conveniently be found through the Mining History Network. The network is based at Exeter University and is available through the website http://info.ex.ac.uk/~RBurt/MinHistNet/welcome.html

8.10.2 British Geological Survey

The BGS produces many useful publications including large and small scale maps, memoirs, reports and atlases on a number of geological, geochemical and geophysical themes. Mineral localities are included in the G-BASE Regional Geochemical Atlases. More details are given in the BGS section (8.1. above).

8.10.3 The Geologist's Directory

Published by the Geological Society, it contains a listing of companies, consultants and contractors active in the earth sciences in Britain together with universities and other educational establishments. It also lists national and local government organisations, including planning authorities. Many sources of information are also given, including libraries, museums and publishers.

8.10.4 Official Yearbook of Britain

Britain 2000. The official yearbook of the United Kingdom (The Stationary Office, 1999) is prepared annually by the Office for National Statistics (Website www.ons.gov.uk). The yearbook describes many features in the life of the country, including the workings of Government and other major institutions. The factual and statistical information in the yearbook is compiled with the co-operation of other Government departments and agencies, and of many other organisations. Sources of more detailed and more topical information (including statistics) are mentioned in the text and a guide to official sources is given in an Appendix.

8.10.5 Mine plans

Plans of abandoned metalliferous and related mines in England and Wales, formerly held by the Health and Safety Executive (HSE), have now been dispersed to the relevant county archive. The Coal Authority holds plans and other information on abandoned coal mines. In Scotland, the BGS holds all the plans of abandoned mines at its Edinburgh office. Some plans are also held for England and Wales in the Keyworth and Exeter Offices.

Health and Safety Executive
Mines Inspectorate
Room 611
Daniel House
Trinity Road
Bootle
Merseyside L20 7HE
Tel: 0151 951 4133
Fax: 0151 951 3758

The Coal Authority
200 Lichfield Lane
Mansfield
Notts NG18 4RG
Tel: 0162 342 7162
Fax: 0162 362 2072

8.10.6 Topographic maps

The whole of Great Britain has been mapped at 1:10 000 scale by the Ordnance Survey with additional coverage at 1:2500 and 1:1250 in most rural and urban areas respectively. Maps are available at a variety of scales and are frequently revised. Digital map data and unpublished information are also available. The Ordnance Survey has an extensive library of aerial photographs of Britain at scales between 1:4000 and 1:26 000. A similar service is provided by the Ordnance Survey of Northern Ireland.

Ordnance Survey
Romsey Road
Southampton
Hampshire SO16 4GU
Tel: 02380 792000
Fax: 02380 792452
Website: www.ordsvy.gov.uk

Ordnance Survey of Northern Ireland
Colby House
Stranmillis Court
Belfast BT9 5BJ
Tel: 02890 255755
Fax: 02890 255700
Website: www.doeni.gov.uk/ordnance/ordnance.htm

8.10.7 Remotely sensed data

Satellite data and other remotely sensed data, and space surveys can be obtained from:

National Remote Sensing Centre
Delta House
Southwood Crescent
Farnborough
Hampshire GU14 0NL
Tel: 0125 254 1464
Fax: 0125 237 5016
Website: www.nrsc.co.uk

Air photographs and other information relating to airborne surveys in Britain is held by the NRSC at:

National Remote Sensing Centre
Arthur Street, Barwell
Leicestershire LE9 8GZ
Tel: 0145 584 9227
Fax: 0145 584 1785
Website: www.nrsc.co.uk

9 FINANCIAL ASSISTANCE

National, regional and local incentives, such as Regional Selective Assistance, are available for industry in certain parts of Britain. Collaborative research programmes can qualify for selective financial support from the DTI under schemes such as EUREKA and LINK, while small companies may be awarded grants under the Smart scheme to help with the development of new products and feasibility studies into innovative technology. Further information and help on financial assistance can be obtained from the Department of Trade and Industry's Invest In Britain Bureau (see above, Section 8.2.1). In addition, regional agencies established to promote local business include:

South-west England:

Devon and Cornwall Development International
2 Derriford Park, Derriford
Plymouth PL6 5QZ
Tel: 01752 793379
Fax: 01752 788660
Website: www.dcdi.co.uk

Wales:

Financial Products Team
Welsh Development Agency
Principality House
The Friary
Cardiff CF1 3XX
Tel: 01443 828718
Fax: 02920 640031
Website: www.wda.co.uk

Mid Wales Division
Ladywell House
Newtown
Powys SY16 1JB
Tel: 01686 613131
Website: www.ruralwales.com/

Scotland:

Scottish Enterprise
120 Bothwell Street
Glasgow G2 7JP
Tel: 0141 248 2700
Fax: 0141 221 3217
Website: www.scotent.co.uk

Northern Ireland:

LEDU (Local Enterprise Development Unit) — the small business agency
LEDU House
Upper Galwally
Belfast BT8 6TB
Tel: 02890 491031
Fax: 02890 691432
Website: www.ledu-ni.gov.uk

Industrial Development Board for Northern Ireland

IDB House
Chichester Street
Belfast BT1 4JX
Tel: 02890 233233
Fax: 02890 545000
Website: www.idbni.co.uk/ie/main.htm

9.1 Highlands and Islands Enterprise

This organisation was established to promote social and economic development in the Highlands and Islands of Scotland. It provides financial support, training for business and assistance for cultural and community projects. It includes 10 local enterprise companies throughout the region. Its predecessor, the Highlands and Islands Development Board, commissioned a number of reports from Robertson Research International Ltd on the occurrence of, and prospects for, metalliferous and industrial minerals in northern Scotland. More recently it commissioned Dames and Moore to produce a report of the mineral potential of the region. The reports can be consulted at the Board's offices and are available for sale.

Highlands and Islands Enterprise
Bridge House
20 Bridge Street
Inverness
Highland IV1 1QR
Tel: 01463 234 171
Fax: 01463 244 469
Website: www.hie.co.uk/menu.html

10 ACKNOWLEDGEMENTS

This is the first revision and update of the original book which was published in 1990. The revision was carried out under the DTI-funded BGS Minerals Programme under the direction of Dr W Hatton, Minerals Group Manager. Numerous members of BGS staff contributed useful information and criticism, notably Dr J S Coats and A G Gunn. D E Highley and Dr G R Chapman updated the section on planning legislation and Mrs S Kimbell and Dr J Cornwell the section on geophysical techniques. The assistance of Mr K Bate, Crown Estate Mineral Agent, who contributed information and advice on the licensing of exploration for precious metals, is greatly appreciated. Mr D G Cameron assisted greatly in preparing the illustrations and cover. Dr H W Haslam provided editorial comment. J I Rayner assisted in the preparation of some of the digital images. This publication was typeset by Mr Adrian Minks under the direction of Mr John Stephenson.

11 REFERENCES

Abrahams, P W and Thornton, I. 1987. Distribution and extent of land contaminated by arsenic and associated metals in mining regions of south-west England. *Transactions Institution of Mining and Metallurgy (Section B: Applied earth science)*, Vol. 96, B1–8.

Al Ajely, K O, Andrews, M J and Fuge, R. 1984. Biogeochemical dispersion patterns associated with porphyry-style mineralisation in the Coed-y-Brenin Forest, North Wales. 1–10 in *Prospecting in areas of glaciated terrain 1984*. [Conference Volume] (London: Institution of Mining and Metallurgy).

Alderton, D H M. 1988. Ag-Au-Te mineralisation in the Ratagain complex, north-west Scotland. *Transactions Institution of Mining and Metallurgy (Section B: Applied earth science)*, Vol. 97, B171–180.

Alderton, D H M. 1993. Mineralization associated with the Cornubian granite batholith. 270–354 in Pattrick, R A D and Polya, D A (editors) *Mineralization in the British Isles*. (London: Chapman and Hall).

Allen, P M. 1980. Copper mineralisation in Great Britain. 266–276 in Jankovic, J and Sillitoe, R H (editors) European copper deposits. *Special Publication Society for Geology Applied to Mineral Deposits,* No. 1 — Belgrade.

Allen, P M and Easterbrook, G D. 1978. Mineralised breccia pipe and other intrusion breccias in the Harlech Dome, North Wales. *Transactions Institution of Mining and Metallurgy (Section B: Applied earth science)*, Vol. 87, B157–161.

Anderson, G M and MacQueen, R W. 1982. Ore Deposit Models 6. Mississippi Valley-type Lead-Zinc deposits. *Geoscience Canada*, Vol. 9, 108–117.

Anderton, R, Bridges, P H, Leeder, M R and Sellwood, B W. 1979. *A dynamic stratigraphy of the British Isles*. (London: George Allen and Unwin).

Andrew, C J, Crowe, R W A, Finlay, S, Pennell, W M and Pyne, J F (editors). 1986. *Geology and Genesis of Mineral Deposits in Ireland* (Dublin: Irish Association for Economic Geology).

Anfield, J, Bent, D, Huxtable, P L, White, C J and Elkins, J D. 1998. Vein mineral extraction and processing in a National Park. 65–76 in *Minerals, Land and the Natural Environment*. Conference Volume, Institution of Mining and Metallurgy, London.

Annels, A E and Roberts, D E. 1989. Turbidite-hosted gold mineralization at the Dolaucothi Gold Mines. Dyfed, Wales, United Kingdom. *Economic Geology*, Vol. 84, 1293–1314.

Anon. 1982. Metal Mining in the UK. *Mining Magazine,* 398–409.

Anon. 1983a. The Mines (Working Facilities & Support) Acts 1966 & 1974 (London: Surveyors Publications).

Anon. 1983b. Legal aspects of prospecting in the United Kingdom. Occasional paper No. 4. (London: Institution of Mining and Metallurgy).

Anon. 1986. Access to mineral resources in Great Britain — the choice. Discussion paper. Royal Institution of Chartered Surveyors GCPPA/Report (86) 6.

Applied Geochemistry Research Group. 1973. Provisional geochemical atlas of Northern Ireland. *Technical Communication*, No. 60. (London: Imperial College).

Appold , M S and Garven, G. 1999. Hydrology of ore formation in the Southeast Missouri District: numerical models of topography-driven fluid flow during the Ouachita orogeny. *Economic Geology*, Vol. 94, 913–936.

Ashcroft, W A and Boyd, R. 1976. The Belhelvie mafic igneous intrusion, Aberdeenshire — a re-investigation. *Scottish Journal of Geology*, Vol. 12, 1–14.

Ashcroft, W A and Munro, M. 1978. The structure of the eastern part of the Insch mafic intrusion, Aberdeenshire. Scottish *Journal of Geology*, Vol. 14, 55–79.

Badham, J P N. 1981. Shale-hosted Pb-Zn deposits: products of exhalation of formation water? *Transactions Institution of Mining and Metallurgy (Section B: Applied earth science),* Vol. 90, B70–76.

Baines, S J, Burley, S D and Gize, A P. 1991. Sulphide mineralisation and hydrocarbon migration in North Sea oilfields. 507–510 in Pagel, M and Leroy, J L (editors). Source, Transport and Deposition of Metals. (Rotterdam: Balkema).

Ball, T K and Bland, D J. 1985. The Cae Coch volcanogenic massive sulphide deposit, Trefriw, North Wales. *Journal Geological Society London*, Vol. 142, 889–898.

Ball, T K and Colman, T B. 1998. Geochemistry of caldera and wallrock alteration associated with volcanogenic sulphide mineralisation, Snowdonia, North Wales. *Transactions Institution of Mining and Metallurgy (Section B: Applied earth science),* Vol. 107, B63–76.

Ball, T K, Basham, I R and Michie, U Mc L. 1982b. Uraniferous vein occurrences of South-West England — Paragenesis and genesis. 113–158 in Vein-type and similar uranium deposits in rocks younger than Proterozoic. (Vienna: International Atomic Energy Agency).

Ball, T K, Basham, I R, Bland, D and Smith, T K. 1982a. Aspects of the geochemistry of bismuth in south-west England. *Proceedings Ussher Society*, Vol. 5, 376–382.

Ball, T K, Cameron, D G, Colman, T B and Roberts, P D. 1991. Behaviour of radon in the geological environment: a review. *Quarterly Journal of Engineering Geology*, Vol. 24, 169–182.

Ball, T K, Fortey, N J and Shepherd, T J. 1985a. Mineralisation at the Carrock Fell tungsten mine, Northern England: paragenetic, fluid inclusion and geochemical study. *Mineralium Deposita*, Vol. 20, 57–65.

Ball, T K, Nicholson, R A and Peachey, D. 1985b. Gas geochemistry as an aid to detection of buried mineral deposits. *Transactions Institution of Mining and Metallurgy (Section B: Applied earth science),* Vol. 94, B181–188.

Bateson, J H, Evans, A D and Johnson, C C. 1984. Investigation of magnetic anomalies and potentially mineralised structures in Whin Sill, Northumberland, England. *Transactions Institution of Mining and Metallurgy (Section B: Applied earth science),* Vol. 93, B71–77.

Beer, K E and Ball, T K. 1986. Tin and tungsten in pelitic rocks from South-west England and their behaviour in contact zones of granites and in mineralised areas. *Proceedings Ussher Society*, Vol. 6, 330–337.

Beer, K E and Ball, T K. 1987. Tungsten mineralisation and magmatism in South-west England. *Chronique Recherche Miniere*, Vol. 487, 53–62.

Beer, K E and Fenning, P J. 1976. Geophysical anomalies and mineralisation at Sourton Tors, Okehampton, Devon. *Report Institute Geological Sciences*, No. 76/1.

Bennett, M A. 1987. Genesis and diagenesis of the Cambrian manganese deposits, Harlech, North Wales. *Geological Journal*, Vol. 22, 7–18.

Berridge, N G. 1969. A summary of the mineral resources of the 'Crofter Counties' of Scotland. *Report Institute Geological Sciences*, No. 69/5.

Beveridge, R, Brown, S, Gallagher, M J and Merritt, J W. 1991. Economic Geology. 545–595 in Geology of Scotland, 3rd Edition. Craig, G Y (editor). (London: Geological Society of London).

Bevins, R E, Alderton, D H M and Horak, J M. 1988. Lead-antimony mineralisation at Bwlch Mine, Deganwy, Wales. *Mineralogical Magazine*, Vol. 52, 391–394.

Bevins, R E. 1985. Pumpellyite-dominated metadomain alteration at Builth Wells, Wales — evidence for a fossil submarine hydrothermal system. *Mineralogical Magazine*, Vol. 49, 451–456.

Bevins, R E. 1994. A mineralogy of Wales. *National Museum of Wales, Geological Series,* No. 16, Cardiff.

Bick, D. 1985. The old copper mines of Snowdonia. (Newent, Glos: The Pound House).

Boast, A M, Harris, M and Steffe, D. 1990. Intrusive-hosted gold mineralization at Hare Hill, Southern Uplands, Scotland. *Transactions of Institution of Mining and Metallurgy (Section B: Applied earth science),* Vol. 99, B106–112.

Bottrell, S H, Shepherd, T J, Yardley, B W D and Dubessy, J. 1988. A fluid inclusion model for the genesis of the ores of the Dolgellau Gold Belt, North Wales. *Journal Geological Society London*, Vol. 145, 139–145.

Bowie, S H U, Ostle, D and Campbell, C B. 1973. Uranium mineralisation in northern Scotland, Wales, the Midlands and south-west England. *Transactions of Institution of Mining and Metallurgy (Section B: Applied earth science),* Vol. 82, B177–179.

Bowker, A M and Hill, I A. 1987. Geophysical study of Kerry Road Orebody, Gairloch. *Transactions of Institution of Mining and Metallurgy (Section B: Applied earth science),* Vol. 96, B213–220.

Bradley, R I, Rudeforth CC and Wilkins, C. 1978. Distribution of some chemical elements in the soils of north-west Pembrokeshire. *Journal of Soil Science*, Vol. 29, 258–270.

Braithwaite, R S W. 1994. Mineralogy of the Alderley Edge — Mottram St Andrew area, Cheshire, England. *Journal of the Russell Society*, Vol. 5, 91–102.

Brammall, D. 1926. Gold and silver in the Dartmoor granite. *Mineralogical Magazine*, Vol. 21, 14–20.

Briskey, J A, Dingess, P R, Smith, F, Gilbert, R C, Armstrong, A K and Cole, G P. 1986. Localisation and source of Mississippi Valley-type zinc deposits in Tennessee, USA, and comparisons with Lower Carboniferous rocks of Ireland. 635–661 in *Geology and Genesis of Mineral Deposits in Ireland.* Andrew, C J, Crowe, R W A, Finlay, S, Pennell, W M and Pyne, J F (editors). (Dublin: Irish Association for Economic Geology).

British Geological Survey. 1996a. Metallogenic Map of Britain and Ireland. Colman, T B, Scrivener, R C, Morris, J H, Long, C B, O'Connor, P J, Stanley, G and Legg, I C (compilers). 1:1.5 M (Keyworth, Nottingham: British Geological Survey).

British Geological Survey. 1996b. Industrial Mineral Resources Map of Britain. Highley, D E, Chapman, G R, Warrington, G and Cameron, D G (compilers). 1:1 M (Keyworth, Nottingham: British Geological Survey).

British Geological Survey. 1996c. Tectonic Map of Britain, Ireland and adjacent areas. Pharaoh, T C, Morris, J H, Long, C B and Ryan, P D (compilers). 1:1.5 M (Keyworth, Nottingham: British Geological Survey).

British Geological Survey. 1997. Regional geochemistry of parts of North-west England and North Wales. (Keyworth, Nottingham: British Geological Survey).

British Geological Survey. 1999a. Coal Resources Map of Britain. Chapman, G R. (Keyworth, Nottingham: British Geological Survey).

British Geological Survey. 1999b. Regional geochemistry of Wales and part of west-central England: stream water. (Keyworth, Nottingham: British Geological Survey).

Buchanan, D L and Dunton, S N (editors). 1996. Precious-metal distribution in Shetland: refinement of targets for gold exploration. (Lerwick: Shetlands Islands Council). 23pp.

Butcher, A R, Pirrie, D, Prichard, H M and Fisher, P. 1999. Platinum-group mineralization in the Rum layered intrusion, Scottish Hebrides, UK. *Journal Geological Society London*, Vol. 156, 213–216.

Butcher, N J D and Hedges, J D. 1987. Exploration and extraction of structurally and lithostratigraphically controlled fluorite deposits in Castleton-Bradwell area of Southern Pennine Orefield, England. *Transactions Institution of Mining and Metallurgy (Section B: Applied earth science),* Vol. 96, B149–155.

Camm, G S and Hosking, K F G. 1985. Stanniferous placer development on an evolving land surface with special reference to placers near St. Austell, Cornwall. *Journal Geological Society London*, Vol. 142, 803–813.

Camm, G S. 1995. Gold in the counties of Cornwall and Devon. (St. Austell, Cornwall: Cornish Hillside Publications).

Carlon, C J. 1979. The Alderley Edge mines. (Altrincham: J. Sherrat and Son).

Chadwick, R A, Holliday, D W, Holloway, S and Hulbert, A G. 1995. The Northumberland-Solway Basin and adjacent areas. Subsurface. *Memoir of the British Geological Survey.*

Charley, M J, Hazleton, R E and Tear, S J. 1989. Precious-metal mineralization associated with Fore Burn igneous complex, Ayrshire, south-west Scotland. *Transactions Institution of Mining and Metallurgy (Section B: Applied earth science),* Vol. 98, B48.

Christoffersen, J E and King, P M. 1988. Hemerdon, an evaluation case history. 27–39 in Extractive Industry Geology 1985. Scott, P W (editor). *Geological Society Miscellaneous Paper,* No. 18.

Clayton, R E, Scrivener, R C and Stanley, C J. 1990. Mineralogical and preliminary fluid inclusion studies of lead-antimony mineralisation in north Cornwall. *Proceedings of the Ussher Society*, Vol. 7, 258–262.

Cliff, D C and Wolfenden, M. 1992. The Lack gold deposits, Northern Ireland. 65–75 87 in Bowden, A A, Earls, G, O'Connor, P G and Pyne, J F (editors). *The Irish Minerals Industry* 1980–1990. (Dublin: Irish Association for Economic Geology).

Clifford, J A, Earls, G, Meldrum, A H and Moore, N. 1992. Gold in the Sperrin Mountains, Northern Ireland: an exploration case history. 77–87 in Bowden, A A, Earls, G, O'Connor, P G and Pyne, J F (editors). *The Irish Minerals Industry 1980–1990.* (Dublin: Irish Association for Economic Geology).

Clifford, J A, Meldrum, A H, Parker, R T G and Earls, G. 1990. 1980–90: a decade of gold exploration in Northern Ireland and Scotland. *Transactions Institution of Mining and Metallurgy (Section B: Applied earth science),* Vol. 99, B133–138.

Clifford, J A. 1986. A note on gold mineralisation in Co. Tyrone. 45–47 in *Geology and Genesis of Mineral Deposits in Ireland.* Andrew, C J, Crow, R W A, Finlay, S, Pennell, W M and Pyne, J F (editors). (Dublin: Irish Association for Economic Geology).

Coats, J S, Fortey, N J, Gallagher, M J and Grout, A. 1984a. Stratiform barium enrichment in the Dalradian of Scotland. *Economic Geology*, Vol. 79, 1585–1595.

Coats, J S, Leake, R C and Peachey, D. 1994. Selective extractions and contrasting sample media in drainage geochemistry. In Hale, M and Plant, J A (editors). Drainage Geochemistry. *Handbook of exploration geochemistry*, Vol. 6. Elsevier, Amsterdam, 601–634.

Coats, J S, Pease, S F and Gallagher, M J. 1984b. Exploration of the Scottish Dalradian. 21–34 in *Prospecting in areas of glaciated terrain 1984* [Conference volume]. (London: Institution of Mining and Metallurgy).

Coats, J S, Smith, C G, Fortey, N J, Gallagher, M J, May, F and McCourt, W J. 1980. Strata-bound barium-zinc mineralization in Dalradian schist near Aberfeldy, Scotland. *Transactions Institution of Mining and Metallurgy (Section B: Applied earth science),* Vol. 89, B110–122.

Collins, R S. 1972. Barium Minerals. *Mineral Dossier, Mineral Resources Consultative Committee*, No. 2.

Collins, R S. 1975. Gold. *Mineral Dossier, Mineral Resources Consultative Committee*, No. 14.

Colman, T B and Cooper, D C. 1998. The combined use of PIMA and Vulcan technology for mineral deposit evaluation based on their application to the Parys Mountain mine, Anglesey. *British Geological Survey Technical Report* WF/98/4C.

Colman, T B. 1982. Titanium distribution in Grampian Region: preliminary report. Unpublished report for client.

Cooper, D C and Thornton, I. 1994. Contaminated terranes. In Hale, M and Plant, J A (editors). Drainage Geochemistry. *Handbook of exploration geochemistry*, Vol. 6. Elsevier, Amsterdam, 447–497.

Cooper, D C, Lee, M K, Fortey, N J, Rundle, C C, Webb, B C and Allen, P M. 1988. The Crummock Water aureole: a zone of metasomatism and source of ore metals in the English Lake District. *Journal of the Geological Society of London*, Vol. 145, 523–540.

Cooper, M P and Stanley, C J. 1990. Minerals of the English Lake District: Caldbeck Fells. (London: Natural History Museum).

Cope, J C W, Ingham, J K and Rawson, P F (editors). 1992. Atlas of Palaeogeography and Lithofacies. *Memoir,* No. 13. (London: Geological Society of London).

Craig, G Y. 1991. Geology of Scotland. (London: Geological Society of London).

Critchley, M F. 1979. A geological outline of the Ecton copper mines, Staffordshire. *Bulletin Peak District Mines Historical Society*, Vol. 7, 177–191.

Crowley, S F, Bottrell, S H, McCarthy, M D B, Ward, J and Young, B. 1997. d34S of Lower Carboniferous anhydrite, Cumbria and its implications for barite mineralization in the northern Pennines. *Journal of the Geological Society of London*, Vol. 154, 597–600.

Crummy, J, Hall, A J, Haszeldine, R S and Anderson, I K. 1997. Potential for epithermal gold mineralization in east and central Sutherland, Scotland: indications from River Brora headwaters. *Transactions Institution of Mining and Metallurgy (Section B: Applied earth science),* Vol. 106, B9–B14.

Curtis, S F, Pattrick, R A D, Jenkin, G R T, Fallick, A E, Boyce, A J and Treagus, J E. 1993. Fluid inclusion and stable isotope study of fault-related mineralization in Tyndrum area, Scotland. *Transactions Institution of Mining and Metallurgy (Section B: Applied earth science),* Vol. 102, B39–48.

Darch, J P and Barber, J. 1983. Multitemporal remote sensing of a geobotanical anomaly. *Economic Geology*, Vol. 78, 770–782.

Darnley, A G, Björklund, A, Bølviken, B, Gustavsson, N, Koval, P V, Plant, J A, Steenfelt, A, Tauchid, M and Xuejing, X. 1995. A global geochemical database for environmental and resource management: recommendations for international geochemical mapping. *Final report of IGCP Project 259.* (Ottawa: UNESCO Publishing).

Dewey, H and Eastwood, T. 1925. Copper ores of the Midlands, Wales, the Lake District and the Isle of Man. Special Report on the Mineral Resources of Great Britain. *Memoir of the Geological Survey of Great Britain*, No. 30.

Dewey, H. 1920. Arsenic and antimony ores. Special Report on the Mineral Resources of Great Britain. *Memoir of the Geological Survey of Great Britain*, No. 15.

Dines, H G. 1956. The metalliferous mining region of South-West England. *Memoir of the Geological Survey of Great Britain.*

Dines, H G. 1958. The West Shropshire mining region. *Bulletin of the Geological Survey of Great Britain*, No. 14, 1–43.

Dominey, S C, Bussell, M A and Camm, G S. 1996. Development of complex, granite-hosted, tin-bearing fracture systems in south-west England: applications of fluid inclusion microfracture studies. *Transactions Institution of Mining and Metallurgy (Section B: Applied earth science),* Vol. 105, B139–143.

Duff, P M D and Smith A J. 1992. Geology of England and Wales. (London: Geological Society of London).

Duller, P R, Gallagher, M J, Hall, A J and Russell, M J. 1997. Glendinning deposit – an example of turbidite-hosted arsenic-antimony-gold mineralization in the Southern Uplands, Scotland. *Transactions Institution of Mining and Metallurgy (Section B: Applied Earth Science),* Vol. 106, B119–B134.

Dunham, K C and Wilson, A A. 1985. Geology of the Northern Pennine Orefield: Volume 2, Stainmore to Craven. *Economic Memoir of the British Geological Survey.*

Dunham, K C. 1952. Fluorspar. 4th edition. Special Report on the Mineral Resources of Great Britain. *Memoir of the Geological Survey of Great Britain*, No. 4.

Dunham, K C. 1983. Ore genesis in the English Pennines: A fluoritic subtype. 86–112 in International Conference on Mississippi Valley-type Lead-Zinc Deposits. [Proceedings volume.] Kisvarsanyi, G, Grant, S K, Pratt, W P and Koenig, J W, (editors). (Rolla: University of Missouri-Rolla).

Dunham, K C. 1990. Geology of the Northern Pennine Orefield: Volume 1, Tyne to Stainmore (2nd edition). *Economic Memoir of the British Geological Survey.*

Dunham, K, Beer, K E, Ellis, R A, Gallagher, M J, Nutt, M J C and Webb, B C. 1979. United Kingdom. 263–317 in *Mineral Deposits of Europe*, Vol. 1: Northwest Europe. Bowie, S H U, Kvalheim, A and Haslam, H W (editors). (London: Institution of Mining and Metallurgy and Mineralogical Society).

Dunlop, A C and Meyer, W T. 1973. Influence of late Miocene — Pliocene submergence on regional distribution of tin in stream sediments, south-west England. *Transactions Institution of Mining and Metallurgy (Section B: Applied earth science),* Vol. 82, B62–64.

Earls, G, Clifford, J A and Meldrum, A H. 1989. Curranghinalt gold deposit, Col Tyrone, Northern Ireland. *Transactions Institution of Mining and Metallurgy (Section B: Applied earth science)*, Vol. 98, B50–51.

Earls, G, Parker, R T G, Clifford, J A and Meldrum, A H. 1992. The geology of the Cononish gold-silver deposit, Grampian Highlands of Scotland. 89–103 in Bowden, A A, Earls, G, O'Connor, P G and Pyne, J F (editors). *The Irish Minerals Industry 1980–1990.* (Dublin: Irish Association for Economic Geology).

Earp, J R. 1958. Mineral veins of the Minera-Maeshafn District of North Wales. *Bulletin of the Geological Survey of Great Britain,* No. 14, 44–69.

Eastwood, T, Hollingsworth, S E, Rose, W C C and Trotter, F M. 1968. Geology of the country around Cockermouth and Caldbeck. *Memoir of the Geological Survey of Great Britain.*

Edmunds, W M. 1971. Hydrogeochemistry of groundwater in the Derbyshire Dome with special reference to trace constituents. *Report Institute of Geological Sciences*, No. 71/7.

Edwards, R P. 1976. Aspects of trace metal and ore distribution in Cornwall. *Transactions Institution of Mining and Metallurgy (Section B: Applied earth science),* Vol. 85, B83–90.

Embrey, P G and Symes, R F. 1987. *Minerals of Cornwall and Devon.* (London: British Museum of (Natural History)).

Emo, G T. 1986. Some considerations regarding the styles of mineralisation at Harberton Bridge, Co. Kildare. 461–470 in *Geology and Genesis of Mineral Deposits in Ireland.* Andrew, C J, Crow, R W A, Finlay, S, Pennell, W M and Pyne, J F (editors). (Dublin: Irish Association for Economic Geology).

Evans, A M (editor). 1982. *Metallization associated with acid magmatism.* (Chichester: Wiley).

Firman, R J and Bagshaw, C. 1974. A re-appraisal of the controls of non-metallic gangue mineral distribution in Derbyshire. *Mercian Geologist*, Vol. 5, 145–161.

Firman, R J. 1978. Epigenetic mineralisation. 226–241 in The geology of the Lake District. Moseley, F, (editor). *Occasional Publication Yorkshire Geological Society* No. 3.

Fletcher, C J N. 1988. Tidal erosion, solution cavitites and exhalative mineralization associated with the Jurassic unconformity at Ogmore, South Glamorgan. *Proceedings Geologists Association*, Vol. 99, 1–14.

Fletcher, C J N, Swainbank, I G and Colman, T B. 1993. Metallogenic evolution in Wales: constraints from lead isotope modelling. *Journal of the Geological Society of London*, Vol. 150, 77–82.

Fletcher, T A and Rice, C M. 1989. Geology, mineralization (Ni-Cu) and precious-metal geochemistry of Caledonian mafic and ultramafic intrusions near Huntly, north-east Scotland. *Transactions Institution of Mining and Metallurgy (Section B: Applied earth science),* Vol. 98, B185–200.

Fletcher, T A, Boyce, A J, Fallick, A E, Rice, C M and Kay, R L F. 1997. Geology and stable isotope study of Arthrath mafic intrusion and Ni-Cu mineralisation, north-east Scotland. *Transactions Institution of Mining and Metallurgy (Section B: Applied earth science),* Vol. 106, B169–178.

Flight, D M A, Hall, G E M and Simpson, P R. 1994. Regional geochemical mapping of Pt, Pd and Au over an obducted ophiolite complex, Shetland Islands, northern Scotland. *Transactions Institution of Mining and Metallurgy (Section B: Applied earth science)*, Vol. 103, B68–78.

Ford, T D and Ineson, P R. 1971. The fluorspar mining potential of the Derbyshire ore field. *Transactions Institution of Mining and Metallurgy (Section B: Applied earth science)*, Vol. 80, B186–210.

Ford, T D, Sergeant, W A S and Smith, M E. 1993. The minerals of the Peak District of Derbyshire. *Bulletin of the Peak District Mines Historical Society*, Vol. 12, 16–55.

Ford, T D. 1976. The ores of the South Pennines and Mendip Hills, England - a comparative study. 161–195 in Handbook of Strata-bound and Stratiform Ore Deposits, II. Regional Studies and Specific Deposits, 5, Regional Studies. Wolf, K H (editor). (Amsterdam: Elsevier).

Fortey, N J, Ingham, J D, Skilton, B R H, Young, B and Shepherd, T J. 1984. Antimony mineralisation at Wet Swine Gill, Caldbeck Fells, Cumbria. *Proceedings Yorkshire Geological Society*, Vol. 45, 59–65.

Fortey, N J and Smith, C G. 1986. Stratabound mineralisation in Dalradian rocks near Tyndrum, Perthshire. *Scottish Journal of Geology*, Vol. 22, 377–393.

Fortey, N J, Coats, J S, Gallagher, M J, Smith, C G and Greenwoood, P G. 1993. New stratabound barite and base-metals in Middle Dalradian rocks near Braemar, north-east Scotland. *Transactions Institution of Mining and Metallurgy (Section B: Applied earth science)*, Vol. 102, B55–64.

Fortey, N J. 1980. Hydrothermal mineralisation associated with minor late Caledonian intrusions in Northern Britain: preliminary comments. *Transactions Institution of Mining and Metallurgy (Section B: Applied earth science)*, Vol. 89, B173–176.

Gallagher, M J, Hall, I H S and Stephenson, D. 1982. Controls and genesis of baryte veins in central Scotland. *Bulletin du B.R.G.M. (2), section II, numero 2*, 143–148.

Gallagher, M J, Michie, U McL, Smith, R T and Haynes, L. 1971. New evidence of uranium and other mineralisation in Scotland. *Transactions Institution of Mining and Metallurgy (Section B: Applied earth science)*, Vol. 80, B150–173.

Gallagher, M J, Smith, R T, Fortey, N J and Parker, M E. 1986. Lead-zinc exploration in the Lower Carboniferous South Scotland. 39–47 in *Prospecting in areas of glaciated terrain 1986*. [Conference volume] (London Institution of Mining and Metallurgy).

Gallagher, M J. 1974. Rutile and zircon in Northumbrian beach sands. *Transactions Institution of Mining and Metallurgy (Section B: Applied earth science)*, Vol. 83, B97–98.

Gawthorpe, R L. 1987. Tectono-sedimentary evolution of the Bowland Basin, N. England, during the Dinantian. *Journal Geological Society London*, Vol. 144, 59–71.

Gayer, R A and Criddle, A J. 1969. Mineralogy and genesis of the Llanharry iron ore deposits. 605–626 in *Proceedings Ninth Commonwealth Mining and Metallurgical Congress*, Vol. 2 (London: Institution of Mining and Metallurgy).

Glentworth, R and Muir, J W. 1963. Soils around Aberdeen, Inverurie and Frazerburgh. Memoir Soil Survey of Great Britain (London: HMSO).

Goldring, D C and Greenwood, D A. 1990. Fluorite mineralization at Beckermet iron ore mine, Cumbria, north England. *Transactions Institution of Mining and Metallurgy (Section B: Applied earth science)*, Vol. 99, B113–119.

Gough, D. 1965. Structural analysis of the ore shoots at Greenside Lead Mine, Cumberland, England. *Economic Geology*, Vol. 60, 1459–77.

Green, G W. 1958. The Central Mendip Lead-Zinc Orefield. *Bulletin of the Geological Survey of Great Britain*, No. 14, 70–90.

Greenbaum, D. 1987. Lithological discrimination in central Snowdonia using airborne multispectral scanner imagery. *International Journal Remote Sensing*, Vol. 8, 799–816.

Greenwood, D A and Smith, F W. 1977. Fluorspar mining in the Northern Pennines. *Transactions Institution of Mining and Metallurgy (Section B: Applied earth science)*, Vol. 86, B181–190.

Gregory, J W. 1928. The nickel-cobalt ore of Talnotry, Kirkudbrightshire. *Transactions Institution of Mining and Metallurgy*, Vol. 37, 178–195.

Groves, A W. 1952. Wartime investigations into the hematite and manganese ore resources of Great Britain and Northern Ireland. *Ministry of Supply Monograph*, No. 20–703.

Gunn, A G, Wiggans, G N, Collins, G L, Rollin, K E and Coats, J S. 1997. Artificial intelligence in mineral exploration: potential applications by SMEs in Britain. *British Geological Survey Technical Report* WF/97/3C.

Gunn, A G. 1989. Drainage and overburden geochemistry in exploration for platinum-group element mineralisation in the Unst ophiolite, Shetland, UK. *Journal of Geochemical Exploration*, Vol. 31, 209–236.

Haggerty, R, Rohl, B M, Budd, P D and Gale, N H. 1996. Pb-isotope evidence on the origin of the West Shropshire orefield, England. *Geological Magazine*, Vol. 133, 611–617.

Haggerty, R. 1995. The mineralization of the Llanwrst orefield, North Wales. UK *Journal of Mines and Minerals*, Vol. 15, 5–10.

Hall, A J. 1993. Stratiform mineralization in the Dalradian of Scotland. 38–101 in Pattrick, R A D and Polya, D A (editors). *Mineralization in the British Isles*. (London: Chapman and Hall).

Hall, G W. 1989. The last 100 years of mining at Pontrhydygroes, Dyfed. *Mining Magazine*, Vol. 116–121.

Hancock, P L (editor). 1983. The Variscan fold belt in the British Isles. (Bristol: Adam Hilger).

Harris, A L, Holland, C H and Leake, B E (editors). 1979. The Caledonides of the British Isles — reviewed. *Special Publication Geological Society London*, No. 8.

Harris, M, Kay, E A, Widnall, M A, Jones, E R and Steele, G B. 1988. Geology and mineralisation of the Lagalochan intrusive complex, Western Argyll, Scotland. *Transactions Institution of Mining and Metallurgy (Section B: Applied earth science)*, Vol. 97, B15–21.

Harris, P M. 1995. The United Kingdom Minerals Industry (2nd Edition). (Keyworth, Nottingham: British Geological Survey).

Harrison, D J. 1985. Mineralogical and chemical appraisal of Corrycharmaig serpentinite intrusion, Glen Lochay, Perthshire. *Transactions Institution of Mining and Metallurgy (Section B: Applied earth science)*, Vol. 94, B147–151.

Harrison, M and Machin, S (Editors). 1999. Mineral and waste planning officers and authorities in Great Britain 1999. (Northallerton: Mineral Planning).

Haslam, H W, Cameron, D G and Evans, A D. 1990. The Mineral Reconnaissance Programme 1990. *British Geological Survey Technical Report* WF/90/6 (BGS Mineral Reconnaissance Report 114).

Hatton, W, Colman, T, Cooper, D, Gunn, A and Plant, J A. 1998. Mineral exploration and information technology. 41–63 in *Minerals, land and the natural environment. Conference volume*. (London: Institution of Mining and Metallurgy).

Hawkes, J R, Harris, P M, Dangerfield, J, Strong, G E, Davis, A E, Nancarrow, P H A, Francis, A D and Smale, C V. 1987. The lithium potential of the St. Austell Granite. *Report British Geological Survey*, Vol. 19, No. 4.

Heddle, M F. 1901. The mineralogy of Scotland. Goodchild, J G (editor). 2 volumes. (Edinburgh: Douglas, D).

Hitzman, M W. 1995. Mineralization in the Irish Zn-Pb-(Ba-Ag) orefield. 25–61 in Irish carbonate-hosted Zn-Pb deposits. Anderson, K, Ashton, J, Earls, G, Hitzman, M and Tear, S (editors). *Society of Economic Geologists Guidebook Series 21*.

Holding, S R. 1992. A survey of the metal mines of Shropshire. 3rd Edition. Shropshire Caving and Mining Club. Account No. 12.

HOLLOWAY, S and CHADWICK, R A. 1984. The IGS Bruton Borehole (Somerset, England) and its regional structural significance. *Proceedings Geologists Association*, Vol. 95, 165–174.

HOLMES, I, CHAMBERS, A D, IXER, R A, TURNER, P and VAUGHAN, D J. 1983. Diagenetic process and the mineralisation in the Triassic of Central England. *Mineralium Deposita*, Vol. 18, 365–377.

HOWE, A C A. 1982. Mineral exploration in the UK. *Mining Magazine*, 358–365.

HUGHES, S J S. 1988. The decline of mining in Mid Wales and prospects of revival. *Rock Bottom*, Vol. 5, 12–14.

HUNTING GEOLOGY AND GEOPHYSICS LTD. 1983. Computer correlation of geological, geochemical and geophysical prospecting data with enhanced satellite imagery. 2 volumes. *Report of Crest programme of the European Economic Community*.

HYSLOP, E K, GILLANDERS, R J, HILL, P G and FAKES, R D. 1999. Rare-earth-bearing minerals fergusonite and gadolinite from the Arran granite. *Scottish Journal of Geology*, Vol. 35, 65–69.

INESON, P R and FORD, T D. 1982. The South Pennine orefield: its genetic theories and eastward extension. *Mercian Geologist*, Vol. 8, 285–303.

INSTITUTE OF GEOLOGICAL SCIENCES. 1984. *List of Open File Reports*. (London: Institute of Geological Sciences).

INSTITUTION OF MINING AND METALLURGY. 1985. High heat production (HHP) granites, hydrothermal circulation and ore genesis [Conference volume]. (London: Institution of Mining and Metallurgy).

IXER, R A and STANLEY, J C. 1996. Siegenite-Bearing assemblages found at the Orme Mine, Llandudno, North Wales. *Mineralogical Magazine*, Vol. 60, 978–982.

IXER, R A and DAVIES, J. 1997. Mineralization at Great Orme copper mines, Llandudno, North Wales, UK. *Journal of Mines and Minerals*, Vol. 17, 7–14.

IXER, R A and STANLEY, C J. 1983. Silver mineralisation at Sark's Hope mine, Sark, Channel Island. *Mineralogical Magazine*, Vol. 47, 539–545.

IXER, R A and VAUGHAN, D J. 1993. Lead-Zinc-Fluorite-Baryte deposits of the Pennines, North Wales and the Mendips. 355–418 in PATTRICK, R A D and POLYA, D A (editors) *Mineralization in the British Isles*. (London: Chapman and Hall).

IXER, R A F, PATTRICK, R A D and STANLEY, C J. 1997. Geology, mineralization and genesis of gold mineralization at Callichar Burn – Urlar Burn, Scotland. *Transactions Institution of Mining and Metallurgy (Section B: Applied earth science)*, Vol. 106, B99–108.

JEFFREY, C A. 1997. Replacement mineralization styles and breccia dome formation at Dirtlow Rake fluorite-barite deposit, Castleton, England. *Transactions Institution of Mining and Metallurgy (Section B: Applied earth science)*, Vol. 106, B15–B23.

JONES, E M, RICE, C M and TWEEDIE, J R. 1987. Lower Proterozoic stratiform sulphide deposits in Loch Maree Group, Gairloch, Northwest Scotland. *Transactions Institution of Mining and Metallurgy (Section B: Applied earth science)*, Vol. 96, B128–140.

JONES, O T. 1922. Lead and zinc. The mining district of North Cardiganshire and West Montgomeryshire. *Special Report on the Mineral Resources of Great Britain, Memoir of the Geological Survey of Great Britain*, No. 20.

KIMBELL, S F, MORGAN, D J R and BALL, T K. 1998. Digitisation of the 1957 airborne radiometric survey of Cornwall. *British Geological Survey Technical Report* WK/98/15.

KING, R J. 1967. The minerals of Leicestershire. *Transactions of the Leicester Literary and Philosophical Society*, Vol. 61, 55–64.

KING, R J. 1968. Mineralisation. 112–117 in The geology of the East Midlands. SYLVESTER-BRADLEY, P B and FORD, T D (editors). (Leicester: Leicester University Press).

KNORRING, O VON and CONDLIFFE, E. 1984. On the occurrence of niobium – tantalum and other rare — element minerals in the Meldon Aplite, Devonshire. *Mineralogical Magazine*, Vol. 48, 443–448.

LAMPLUGH, G W. 1903. Economic geology of the Isle of Man, with special reference to the metalliferous mines. *Memoir Geological Survey of Great Britain*. (London: HMSO).

LARGE, D E. 1983. Sediment-hosted massive sulphide lead-zinc deposits: an empirical model. 1–29 in Short course in sediment-hosted stratiform lead-zinc deposits. SANGSTER, D F (editor). *Mineralogical Association of Canada*.

LEACH, D L and ROWAN, E L. 1986. Genetic link between Ouachita foldbelt tectonism and the Mississippi Valley-type lead-zinc deposits of the Ozarks. *Geology*, Vol. 14, 931–935.

LEAKE, B C. 1982. Volcanism in the Dalradian. 45–50 in Igneous Rocks of the British Isles. SUTHERLAND, D S (editor). (Chichester: Wiley).

LEAKE, R C and AUCOTT, J W. 1973. Geochemical mapping and prospecting by use of rapid automatic X-ray fluorescence analysis of panned concentrates. 389–400 in Geochemical exploration 1972. JONES, M J (editor). (London: Institution of Mining and Metallurgy).

LEAKE, R C and STYLES, M T. 1984. Borehole sections through the Traboe hornblende schists, a cumulate complex overlying the Lizard peridotite. *Journal Geological Society London*, Vol. 141, 41–52.

LEAKE, R C, BLAND, D J and COOPER, C. 1993. Source characterisation of alluvial gold from mineral inclusions and internal compositional variation. *Transactions Institution of Mining and Metallurgy (Section B: Applied earth science)*, Vol. 102, B65–82.

LEAKE, R C, BLAND, D J, STYLES, M T and CAMERON, D G. 1991. Internal structure of Au-Pd-Pt grains from south Devon, England, in relation to low-temperature transport and deposition. *Transactions Institution of Mining and Metallurgy (Section B: Applied earth science)*, Vol. 100, B159–178.

LEAKE, R C, CHAPMAN, R J, BLAND, D J, CONDLIFFE, E and STYLES, M T. 1997. Microchemical characterisation of alluvial gold from Scotland. *Transactions Institution of Mining and Metallurgy (Section B: Applied earth science)*, Vol. 106, B85–98.

LEAKE, R C, STYLES, M T, BLAND, D J, HENNEY, P J, WETTON, P D and NADEN, J. 1995. The interpretation of alluvial gold characteristics as an exploration technique. *British Geological Survey Technical Report* WC/95/22.

LEAKE, R C, CHAPMAN, R J, BLAND, D J, STONE, P, CAMERON, D G and STYLES, M T. 1998. The origin of alluvial gold in the Leadhills area of Scotland: evidence from interpretation of internal chemical characteristics. *Journal of Geochemical Exploration*, Vol. 63, 7–36.

LEGG, I C, PYNE, J F, NOLAN, C, MCARDLE, P, FLEGG, A and O'CONNOR, P J. 1985. Mineral localities in the Dalradian and associated igneous rocks of County Donegal, Republic of Ireland and of Northern Ireland. *Geological Survey of Ireland, Report Series* RS 85/3 (Mineral Resources).

LEYSHON, P R and CAZELET, P C D. 1976. Base-metal exploration programme in Lower Palaeozoic volcanic rocks, Co. Tyrone, Northern Ireland. *Transactions Institution of Mining and Metallurgy (Section B: Applied earth science)*, Vol. 85, B91–99.

LORD, R A and PRICHARD, H M. 1997. Exploration and origin of stratigraphically controlled platinum-group element mineralization in crustal-sequence ultramafics, Shetland ophiolite complex. *Transactions Institution of Mining and Metallurgy (Section B: Applied earth science)*, Vol. 106, B179–193.

LORD, R A, PRICHARD, H M and NEARY, C R. 1994. Magmatic platinum-group element concentrations and hydrothermal upgrading in Shetland ophiolite complex. *Transactions Institution of Mining and Metallurgy (Section B: Applied earth science)*, Vol. 103, B87–106.

LOWRY, D, BOYCE, A J, FALLICK, A E and STEPHENS, W E. 1997. Sources of sulphur, metals and fluids in granitoid-related mineralization of the Southern Uplands, Scotland. *Transactions Institution of Mining and Metallurgy (Section B: Applied earth science)*, Vol. 106, B157–168.

LYDON, J W. 1986. Metalliferous hydrothermal systems in sedimentary rocks. 555–578 in *Geology and Genesis of Mineral*

Deposits in Ireland. Andrew, C J, Crow, R W A, Finlay, S, Pennell, W M and Pyne, J F (editors). (Dublin: Irish Association for Economic Geology).

Lydon, J W. 1988. Ore Deposit Models # 14, Volcanogenic Massive Sulphide Deposits Part 2: *Genetic Models. Geoscience Canada*, Vol. 15, 43–65.

Macklin, M G and Dowsett, R B. 1989. The chemical and physical speciation of trace metals in fine grained overbank flood sediments in the Tyne basin, North-east England. *Catena*, Vol. 16, 135–151.

Mann, A W, Birrell, R D, Mann, A T, Humphreys, D B and Perdrix, J L. 1998. Application of the mobile metal ion technique to routine geochemical exploration. *Journal of Geochemical Exploration*, Vol. 61, 87–102.

Masheder, R and Rankin, A H. 1988. Fluid inclusion studies on the Ecton Hill copper deposits, north Staffordshire. *Mineralogical Magazine*, Vol. 52, 473–482.

Mason, J S. 1997. Regional polyphase and polymetallic vein mineralization in the Caledonides of the Central Wales Orefield. *Transactions Institution of Mining and Metallurgy (Section B: Applied Earth Science)*, Vol. 106, B135–B143.

McArdle, P. 1990. A review of carbonate-hosted base-metal-barite deposits in the Lower Carboniferous rocks of Ireland. *Chronique recherche miniere*, Vol. 500, 3–29.

McCready, A J and Parnell, J. 1998. A Phanerozoic analogue for Witwatersrand-type uranium-titanium-bitumen nodules in Devonian conglomerate/sandstone, Orkney, Scotland. *Transactions Institution of Mining and Metallurgy (Section B: Applied earth science)*, Vol. 107, B89–98.

Michie, U McL and Cooper, D C. 1979. Uranium in the Old Red Sandstone of Orkney. *Report Institute Geological Sciences*, No. 78/16.

Michie, U McL and Gallagher, M J. 1979. Uranium concentration of potential economic significance in Scotland and their genesis. *Open File Report,* No. 336. (London: Institute of Geological Sciences).

Miller, J M and Taylor, K. 1966. Uranium mineralization near Dalbeattie, Kirkcudbrightshire. *Bulletin Geological Survey Great Britain*, Vol. 25, 1–18.

Miller, O D W. 1993. Precious metal mineralization associated with the Coed-y-Brenin porphyry copper system, North Wales.. Unpublished PhD thesis, University of Aberdeen.

Millward, D, Beddoe-Stevens, B and Young, B. 1999. Pre-Acadian copper mineralization in the English Lake District. *Geological Magazine*, Vol. 136, 159–176.

Moore, J McM and Camm, S. 1982. Interactive enhancement of Landsat imagery for structural mapping in tin-tungsten prospecting: a case history of the South West England orefield (UK). 727–740 in *Proceedings of the international symposium on remote sensing of environment, Second Thematic Conference, Fort Worth, Texas*. (Michigan: Environmental Research Institute of Michigan).

Moreton, S. 1996. The Alva silver mine. *The Mineralogical Record*, Vol. 27, 405–414.

Munro, M. 1986. Mylonite zones in the Insch 'Younger Basic' mass. *Scottish Journal of Geology*, Vol. 22, 132–136.

Naden, J and Caulfield, J B D. 1989. Fluid inclusion and isotopic studies of gold mineralisation in the Southern Uplands of Scotland. *Transactions Institution of Mining and Metallurgy (Section B: Applied earth science)*, Vol. 98, B46–48.

Naylor, H, Turner, P, Vaughan, D J, Boyce, A J and Fallick, A E. 1989. Genetic studies of red-bed mineralization in the Triassic of the Cheshire Basin, north-west England. *Journal of the Geological Society of London*, Vol. 146, 685–699.

Newall, P S and Newall, G C. 1989. Use of lithogeochemistry as an exploration tool at Redmoor sheeted-vein complex, east Cornwall, south-west England. *Transactions Institution of Mining and Metallurgy (Section B: Applied earth science)*, Vol. 98, B162–174.

Nicholson, K and Anderton, R. 1989. The Dalradian rocks of the Lecht, NE Scotland: stratigraphy, faulting, geochemistry and mineralisation. *Transactions of the Royal Society of Edinburgh: Earth Sciences*, Vol. 80, 143–157.

Nicholson, K. 1987. Ironstone-Gossan discrimination: pitfalls of a simple geochemical approach - a case study from north-east Scotland. *Journal Geochemical Exploration*, Vol. 27, 239–257.

Nicholson, K. 1990. Stratiform manganese mineralisation near Inverness Scotland: a Devonian lacustrine hot-spring deposit? *Mineralium Deposita*, Vol. 25, 126–131.

Nickless, E F D, Booth, S J and Mosley, P N. 1976. The celestite resources of the area north-east of Bristol. *Mineral Assessment Report,* Vol. 25, Institute of Geological Sciences.

Notholt, A J G, Highley, D E and Harding, P R. 1985. Investigation of phosphate (apatite) potential of Loch Borrolan igneous complex, Northwest Highlands, Scotland. *Transactions Institution of Mining and Metallurgy (Section B: Applied earth science)*, Vol. 94, B58–65.

Panteleyev, A. 1986. Ore Deposits # 10. A Canadian Cordilleran Model for Epithermal Gold-Silver Deposits. *Geoscience Canada*, Vol. 13, 101–111.

Parker, R T G, Clifford, J A and Meldrum, A H. 1989. Cononish gold-silver deposit, Perthshire, Scotland. *Transactions Institution of Mining and Metallurgy (Section B: Applied earth science)*, Vol. 98, B51–52.

Pattrick, R A D and Polya, D. 1993. *Mineralization in the British Isles*. (London: Chapman and Hall).

Pattrick, R A D, Boyce, A and MacIntyre, R M. 1988. Gold-silver vein mineralisation at Tyndrum, Scotland. *Mineralogy and Petrology*, Vol. 38, 61–76.

Pattrick, R A D. 1984. Sulphide mineralogy of the Tomnadashan copper deposit and the Corrie Buie lead veins, South Loch Tayside, Scotland. *Mineralogical Magazine*, Vol. 48, 85–91.

Peach, B N, Horne, J, Gunn, W, Clough, C T and Hinxman, L W. 1907. The geological structure of the North West Highlands of Scotland. *Memoir Geological Survey of Great Britain*. (Glasgow: HMSO).

Peach, B N, Wilson, J S G, Hill, J B, Bailey, E B and Graham, G W. 1911. Explanation of Sheet 28. The Geology of Knapdale, Jura and North Kintyre. *Memoir Geological Survey Scotland, Edinburgh*.

Peacock, J D, Berridge, N G, Harris, A L and May, F. 1968. Explanation of Sheet 95. The geology of the Elgin district. *Memoir Geological Survey Scotland, Edinburgh*.

Pease, S F, Coats, J S and Fortey, N J. 1986. Exploration for sediment-hosted exhalative mineralisation in the Middle Dalradian at Glenshee, Scotland. 94–107 in *Prospecting in areas of glaciated terrain 1986* [Conference volume.] (London: Institution of Mining and Metallurgy).

Petts, G E, Coats, J S and Hughes, N. 1991. Freeze-sampling method of collecting drainage sediments for gold exploration. *Transactions Institution of Mining and Metallurgy (Section B: Applied earth science)*, Vol. 100, B28–32.

Pharaoh, T C, Merriman, R J, Webb, P C and Beckinsale, R D. 1987. The concealed Caledonides of eastern England: preliminary results of a multidisciplinary study. *Proceedings Yorkshire Geological Society*, Vol. 46, 355–369.

Phillips, W J. 1986. Hydraulic fracturing effects in the formation of mineral deposits. *Transactions Institution of Mining and Metallurgy (Section B: Applied earth science)*, Vol. 95, B17–24.

Plant, J A and 29 others. 1998. Multidataset analysis for the development of gold exploration models in western Europe. *British Geological Survey Report*, SF/98/1.

Plant, J A and Jones, D G (editors). 1989. Metallogenic models and exploration criteria for buried carbonate-hosted ore deposits: a multidisciplinary study in eastern England. (London: Institution of Mining and Metallurgy; Keyworth: British Geological Survey).

PLANT, J A, and JONES, D G. 1999. Development of regional exploration criteria for buried carbonate-hosted mineral deposits: A multidisciplinary study in northern England. (Keyworth, Nottingham: the British Geological Survey).

PLANT, J A, JONES, D G, and HASLAM, H W (editors). 2000. The Cheshire Basin. Basin evolution, fluid movement and mineral resources in a Permo-Triassic rift setting. (Keyworth, Nottingham: the British Geological Survey).

PLANT, J A, JONES, D G, BROWN, G C, COLMAN, T B, CORNWELL, J D, SMITH, K, SMITH, N J P, WALKER, A S D and WEBB, P C. 1988. Metallogenic models and exploration criteria for buried carbonate-hosted ore deposits: results of a multidisciplinary study in eastern England. 321–352 in *Mineral deposits within the European Community*. BOISSONAS, J and OMENETTO, P (editors). (Berlin: Springer-Verlag).

PLANT, J A, SIMPSON, P R, GREEN, P M, WATSON, J V and FOWLER, M B. 1983. Metalliferous and mineralised Caledonian granites in relation to regional metamorphism and fracture systems in northern Scotland. *Transactions Institution of Mining and Metallurgy (Section B: Applied earth science)*, Vol. 92, B33–42.

PLANT, J A, SMITH, R T, STEVENSON, A G, FORREST, M D and HODGSON, J F. 1984. Regional geochemical mapping for mineral exploration in northern Scotland. 103–120 in *Prospecting in areas of glaciated terrain 1984* [Conference volume.] (London: Institution of Mining and Metallurgy).

PLANT, J A, STONE, P, FLIGHT, D M A, GREEN, P M and SIMPSON, P R. 1997. Geochemistry of the British Caledonides: the setting for metallogeny. *Transactions Institution of Mining and Metallurgy (Section B: Applied earth science)*, Vol. 106, B67–78.

PLANT, J A. 1986. Models for granites and their mineralising systems in the British and Irish Caledonides. 121–156 in *Geology and genesis of mineral deposits in Ireland*. ANDREW, C J, CROWE, R W A, FINLAY, S, PENNELL, W M and PYNE, J F (editors). (Dublin: Irish Association for Economic Geology).

POINTON, C R and IXER, R A. 1980. Parys Mountain mineral deposit, Anglesey, Wales: geology and ore mineralogy. *Transactions Institution of Mining and Metallurgy (Section B: Applied earth science)*, Vol. 89, B143–155.

POLYA, D A, PATTRICK, R A D, MARSH, D J and BAO, J-J. 1995. Silver and Base Metal Dispersion in Stream Sediments and Waters around an Epithermal Ag-Au-Cu Prospect at Lagalochan, Western Scotland. *Exploration and Mining Geology*, Vol. 4, 271–284.

PRICHARD, H M and LORD, R A. 1988. The Shetland Ophiolite: Evidence for a Supra-Subduction zone origin and implications for Platinum-Group Element Mineralisation. 289–302 in *Mineral Deposits within the European Community*. BOISSONNAS, J and OMENETTO, P (editors). (Berlin: Springer-Verlag).

RAYBOLD, J G. 1974. Ore textures, paragenesis and zoning in the lead-zinc veins of mid-Wales. *Transactions Institution of Mining and Metallurgy (Section B: Applied earth science)*, Vol. 83, B112–119.

READ, D, COOPER, D C and MCARTHUR, J M. 1987. The composition and distribution of nodular monazite in the Lower Palaeozoic rocks of Great Britain. *Mineralogical Magazine*, Vol. 51, 271–280.

REEDMAN, A J, COLMAN, T B, CAMPBELL, S D G and HOWELLS, M F. 1985. Volcanogenic mineralisation related to the Snowdon Volcanic Group (Ordovician), Gwynedd, North Wales. *Journal Geological Society London*, Vol. 142, 875–888.

RICE, C M, ASHCROFT, W A, BATTEN, D J, BOYCE, A J, CAULFIELD, J B D, FALLICK, A E, HOLE, M J, JONES, E, PEARSON, M J, ROGERS, G, SAXTON, J M, STUART, F M, TREWIN, N H and TURNER, G. 1995. A Devonian auriferous hot spring system, Rhynie, Scotland. *Journal of the Geological Society, London*, Vol. 152, 229–250.

RICE, C M. 1993. Mineralization associated with Caledonian intrusive activity. 102–186 in PATTRICK, R A D and POLYA, D A (editors). *Mineralization in the British Isles*. (London: Chapman and Hall).

RICE, R and SHARP, G J. 1976. Copper mineralisation in the forest of Coed-y-Brenin, North Wales. *Transactions Institution of Mining and Metallurgy (Section B: Applied earth science)*, Vol. 85, B1–13.

RICE, R. 1975. Geochemical exploration in an area of glacial overburden at Arthrath, Aberdeenshire. 82–86 in *Exploration in areas of glaciated terrain 1975* [Conference volume]. JONES, M J (editor). (London: Institution of Mining and Metallurgy).

RICHARDSON, J B. 1974. *Metal mining*. (London: Allan Lane).

RIDGWAY, J M. 1983. Silver. Mineral Dossier. *Mineral Resources Consultative Committee*, No. 25.

ROBEY, J A and PORTER, L. 1972. The copper and lead mines of Ecton Hill, Staffordshire. *Moorland Publishing*.

ROSE , W C C and DUNHAM, K C. 1977. Geology and hematite deposits of South Cumbria. *Economic Memoir Geological Survey of Great Britain*, Sheets 58, part 48.

RUSSELL, M J. 1986. Extension and convection: a genetic model for the Irish Carboniferous base-metal and barite deposits. 545–554 in *Geology and genesis of mineral deposits in Ireland*. ANDREW, C J, CROWE, R W A, FINLAY, S, PENNELL, W M and PYNE, J F (editors). (Dublin: Irish Association for Economic Geology).

SANGSTER, D F. 1990. Mississippi Valley-type and sedex lead-zinc deposits: a comparative examination. *Transactions Institution of Mining and Metallurgy (Section B: Applied earth science)*, Vol. 99, B21–42.

SCOTT, B. 1967. Barytes mineralisation at Gasswater mine, Ayrshire, Scotland. *Transactions Institution of Mining and Metallurgy (Section B: Applied earth science)*, Vol. 76, B40–51.

SCOTT, B. 1976. Zinc and lead mineralization along the margins of the Caledonian orogen. *Transactions Institution of Mining and Metallurgy (Section B: Applied earth science)*, Vol. 85, B200–204.

SCOTT, R A, POLYA, D A and PATTRICK, R A D. 1988. Proximal Cu + Zn exhalites in the Argyll Group Dalradian, Creag Bhocan, Perthshire. *Scottish Journal Geology*, Vol. 24, 97–112.

SHEPHERD, T J and ALLEN, P M. 1985. Metallogenesis in the Harlech Dome, North Wales: a fluid inclusion interpretation. *Mineralium Deposita*, Vol. 20, 159–168.

SHEPHERD, T J and BOTTRELL, S H. 1993. Dolgellau Gold Belt, Harlech district, North Wales. 187–207 in PATTRICK, R A D and POLYA, D A (editors). *Mineralization in the British Isles*. (London: Chapman and Hall).

SHEPHERD, T J and GOLDRING, D C. 1993. Cumbrian hematite deposits, North-west England. 419–445 in PATTRICK, R A D and POLYA, D A (editors). *Mineralization in the British Isles*. (London: Chapman and Hall).

SIBLY, T F. 1927. Iron ores — the haematites of the Forest of Dean and South Wales. Special Report on the Mineral Resources of Great Britain. *Memoir of the Geological Survey of Great Britain*, No. 10.

SILLITOE, R H. 1993. Gold-rich porphyry copper deposits: geological model and exploration implications. 465–478 in Mineral deposit modeling. KIRKHAM, R V, SINLCAIR, W D, THORPE, R I and DUKE, J M (editors). *Geological Association of Canada Special*, Paper 40.

SLATER, D and HIGHLEY, D E. 1976. The iron ore deposits in the United Kingdom of Great Britain and Northern Ireland. 393–409 in ZITZMANN A (editor). *The Iron Ore Deposits of Europe and adjacent areas*, Vol. 1. (Hannover: Bundesanhalt fur Geowissenschaften und Rohrstoffe).

SLATER, D. 1973. Tungsten. Mineral Dossier. *Mineral Resources Consultative Committee*, No. 5.

SLATER, D. 1974. Tin. Mineral Dossier. *Mineral Resources Consultative Committee*, No. 9.

SMITH, C G. 1985. Recent investigations of manganese mineralisation in the Scottish Highlands. *Transactions Institution of Mining and Metallurgy (Section B: Applied earth science)*, Vol. 94, B159–162.

SMITH, C G, GALLAGHER, M J, COATS, J S and PARKER, M E. 1984. Detection and general characteristics of strata-bound mineralisation in the Dalradian of Scotland. *Transactions Institution of*

Mining and Metallurgy (Section B: Applied earth science), Vol. 93, B125–133.

Smith, F E. 1988. Fluorspar mining in a national park. *Land and Minerals Surveying*, Vol. 6, 231–241.

Smith, K and Leake, R C. 1984. Geochemical soil surveys as an aid to mapping and interpretation of the Lizard Complex. *Journal Geological Society London*, Vol. 141, 71–78.

Smith, T K and Ball, T K. 1982. XRF analysis of vegetation samples and its application to mineral exploration. *Advances in X-ray analysis*, Vol. 26, 409–414. (New York: Plenum Press).

Stanley, C J and Vaughan, D J. 1982. Copper, lead, zinc and cobalt mineralisation in the English Lake District: classification, condition of formation and genesis. *Journal Geological Society London*, Vol. 139, 569–579.

Stanley, C J, Symes, R F and Jones, G C. 1987. Nickel-copper mineralisation at Talnotry, Newton Stewart, Scotland. *Mineralogy and Petrology*, Vol. 37, 293–313.

Stillman, C J. 1982. Lower Palaeozoic volcanism in Ireland. 113–126 in *Igneous rocks of the British Isles*. Sutherland, D S (editor). (Chichester: Wiley).

Stone, P, Cook, J M, McDermott, C, Robinson, J J and Simpson, P R. 1995. Lithostratigraphic and structural controls on distribution of As and Au in south-west Southern Uplands, Scotland. *Transactions Institution of Mining and Metallurgy (Section B: Applied earth science)*, Vol. 89, B111–119.

Stone, P, Gunn, A G, Coats, J S and Carruthers, R M. 1986. Mineral exploration in the Ballantrae Complex, south-west Scotland. 265–278 in *Metallogeny of basic and ultrabasic rocks* [Conference volume.]. Gallagher, M J (editor). (London: Institution of Mining and Metallurgy).

Strahan, A, Gibson, W, Cantrill, T C, Sherlock, R L and Dewey, H. 1920. Iron Ores (contd.). Pre-Carboniferous and Carboniferous bedded ores of England and Wales. Special Report on the Mineral Resources of Great Britain. *Memoir of the Geological Survey of Great Britain*, No. 13.

Sutherland, D S. 1982. Alkaline intrusions of north-western Scotland. 203–214 in *Igneous Rocks of the British Isles*. Sutherland, D S (editor). (Chichester: Wiley).

Tennant, S C and Steed, G M. 1997. Role of lithogeochemistry in reassessment of the geological setting of Parys Mountain polymetallic sulphide deposit, Anglesey, Wales. *Transactions Institution of Mining and Metallurgy (Section B: Applied earth science)*, Vol. 106, B144–156.

The Stationary Office. 1999. Britain 2000: the official yearbook of the United Kingdom. (London: The Stationary Office).

Thomas, I A. 1973. Celestite. Mineral Dossier. *Mineral Resources Consultative Committee*, No. 6.

Trythall, R J B, Eccles, C, Molyneaux, S G and Taylor W E G. 1987. Age and controls of ironstone deposition (Ordovician) North Wales. *Geological Journal*, Vol. 22, 31–43.

Tylecote, R F. 1986. The prehistory of metallogeny in the British Isles. (London: The Institute of Metals).

Wadsworth, W J. 1982. The basic plutons. 135–148 in *Igneous Rocks of the British Isles*. Sutherland, D S (editor). (Chichester: Wiley).

Warrington, G. 1966. The metalliferous mining district of Alderley Edge, Cheshire. *Mercian Geologist*, Vol. 1, 111–129.

Webb, A. 1980. *Mineral Map of Great Britain*. (Epsom: Nexus Geological Systems).

Webb, P C, Tindle, A G and Ixer, R A. 1992. W-Sn-Mo-Bi-Ag mineralization associated with zinnwaldite-bearing granite from Glen Gairn, Scotland. *Transactions Institution of Mining and Metallurgy (Section B: Applied earth science)*, Vol. 101, B59–72.

Wightman, A. 1996. *Who owns Scotland*. (Edinburgh: Canongate Books).

Williams, C J. 1993. A history of Great Orme Mines. *British Mining*, Vol. 52.

Williamson, J P. 2000. Cultural noise removal from HiRES-1 aeromagnetic data using the Generic Mapping Tools (GMT) package. *British Geological Survey Technical Report* 00/1.

Wilson, G V and Flett, J S. 1921. The lead, zinc, copper and nickel ores of Scotland. Special Report on the Mineral Resources of Great Britain. *Memoir of the Geological Survey of Great Britain*, No. 17.

Windley, B F. 1982. Igneous rocks of the Lewisian complex. 9–18 in *Igneous Rocks of the British Isles*. Sutherland, D S (editor). (Chichester: Wiley).

Windley, B F. 1984. *The Evolving Continents*. (Chichester: Wiley).

Wolfson. 1978. The Wolfson geochemical atlas of England and Wales. (Oxford: Clarendon Press).

Woodland, A W. 1956. The manganese deposits of Great Britain. 197–218 in *Symposium Sobre Yacimientos de Manganeso, Tomo*. Vol. 5. Reyna, J G (editor). 20th International Geological Congress, Mexico.

Young, B, Cooper, D C, Colman, T B and Millward, D. 1992. Barium and base-metal mineralisation associated with the southern margin of the Solway-Northumberland Trough. In Foster, R P (editor). Mineral deposit modelling in relation to crustal reservoirs of the ore-forming elements, abstracts volume. *Institution of Mining and Metallurgy*.

Young, B. 1987. Glossary of the minerals of the Lake District and adjoining area. (Newcastle upon Tyne: British Geological Survey).

Young, T. 1993. Sedimentary iron ores. 446–483 in Pattrick, R A D and Polya, D A (editors). *Mineralization in the British Isles*. (London: Chapman and Hall).

Appendix 1 Known mineralisation by commodity

This section briefly describes the known metalliferous mineral deposits of Britain and provides references for additional reading. It also includes information on mineral exploration in some areas, especially that derived from the activities of the MRP or MEIGA projects. Information on private-sector activities tends to be sporadic and incomplete as there is currently no requirement for companies to deposit data from their mineral exploration programmes with any central agency, except in Northern Ireland. Figure 21 shows the location of the most important orefields and individual deposits referred to in the text.

Antimony

South-west England

Small scale production of antimony has been recorded from the south-west England mining field in north-south veins (locally termed crosscourses) on the fringes of the main tin and copper mining districts (Dines, 1956). The main area of production was around Wadebridge and Port Isaac in north Cornwall with a recorded output of about 250 tons from a number of mines which are listed in Dewey (1920). Antimony occurs in jamesonite ($Pb_4FeSb_6S_{14}$), stibnite (Sb_2S_3) and bournonite ($PbCuSbS_3$) with associated galena and occasional traces of gold in north–south striking quartz-siderite veins which are often associated with Devonian spilitic pillow lavas (Edwards, 1976). The BGS Mineral Reconnaissance Programme (MRP) has carried out geochemical and geophysical surveys in the area and drilled a number of shallow boreholes (Leake et al., 1989). Minor gold-antimony mineralisation occurs in a quartz-carbonate vein near Lodiswell in south Devon (Stanley et al., 1990). The antimony minerals include tetrahedrite ($(Cu, Fe, Zn, Ag)_{12}Sb_4S_{13}$), bournonite and jamesonite. The vein is associated with a north-west–south-east fracture system cutting the Devonian sediments and interbedded basic volcanic rocks. 'Antimonite' (stibnite) and 'fahlerz' (tetrahedrite) occur in the Combe Martin lead-silver mines in North Devon (Dewey, 1921). Museum-quality specimens of bournonite were recovered from the Herodsfoot mine near Liskeard in south Cornwall (Embrey and Symes, 1987).

Scotland

Antimony also occurs in Silurian greywackes at the Louisa mine at Glendinning in south-west Scotland (Duller et al., 1997). The mine produced about 200 t of antimony metal between 1793 and 1891 from a north-north-east striking quartz-stibnite (Sb_2S_3) vein. The Fountainhead mine, near the Knipe on Hare Hill near New Cumnock in south-west Scotland, produced minor amounts of antimony from a north-south striking quartz-stibnite vein cutting a Caledonian granodiorite. The area has been investigated for antimony (MEG 135) and for gold (MEIGA 257). Gold is associated with disseminated arsenopyrite mineralisation within zones of sericitised granodiorite. The highest gold grades occur adjacent to late-stage Sb-Pb veins (Boast et al., 1990). Minor amounts of tetrahedrite have been reported from MRP boreholes in the Black Stockarton Moor porphyry-style mineralisation (Lowry et al., 1997). Antimony is recorded in analyses for gold from Stronchullin in Knapdale, Strathclyde (Peach et al., 1911). More recent gold exploration in the area showed elevated levels of Sb in rock and sediment samples, but did not locate any discrete antimony minerals (Gunn et al., 1997). Tetrahedrite also occurs in the Tomnadashan copper deposit in central Scotland in a porphyry-style association (Pattrick, 1984).

Wales

Stibnite is associated with galena and other antimony minerals veins cutting Ordovician acid ash-flow tuffs at the Bwlch mine near Conwy, North Wales (Bevins et al., 1988). Small amounts of boulangerite ($Pb_5Sb_4S_{11}$), tetrahedrite and bournonite occur in the 'early complex' or 'A1' (Mason, 1997) polymetallic Pb-Zn veins of the Central Wales orefield, but no production of antimony has been reported.

Lake District and the Isle of Man

Stanley and Vaughan (1982) describe minor occurrences of antimony minerals associated with quartz-galena-sphalerite and quartz-stibnite veins in the Lake District. A small quartz-stibnite vein near the Carrock Fell tungsten mine in the northern Lake District is described by Fortey et al. (1984). A comprehensive list of mineral localities in the Lake District, including those containing antimony minerals, is given in Young (1987). 'Plumosite' (stibnite) and tetrahedrite have also been recorded from the major Foxdale Lode on the Isle of Man (Lamplugh, 1903) and from a trial at its western extension in Niarbyl Bay. Lamplugh noted that where it occurred the galena contained an unusual amount of silver (in excess of 100 oz per ton) which may have been due to the admixture of tetrahedrite. Additional information on antimony mining in Britain is given in Dewey (1920).

Arsenic

South-west England

An estimated total of 250 000 tonnes of white arsenic (As_2O_3) was produced in south-west England from veins containing arsenopyrite in association with copper and iron sulphides (Dunham et al., 1979). The main producer was the Devon Great Consols mine near Plymouth with a recorded output of over 70 000 tons between 1848 and 1909 (Dines, 1956). In the 1870s half the world's production was estimated to come from half a dozen mines in the Callington and Tavistock area, including Devon Great Consols. Other centres of arsenic production included the Cambourne–Redruth area where the South Crofty, Dolcoath and East Pool mines all produced significant amounts and the Levant mine on the Land' End peninsula (Dewey, 1920). The arsenopyrite tended to decrease in depth. The smelting of arsenical ores to produce white arsenic in the nineteenth century contaminated large areas downwind of smelters in the region.

Lake District

Arsenopyrite occurs in many of the Lake District mineral veins, especially at the Carrock Fell tungsten deposit and in the Coniston copper mines (Stanley and Vaughan, 1982). A comprehensive list of mineral localities in the Lake District, including those containing arsenic minerals, is given in Young (1987).

Wales

Arsenopyrite is an important constituent of the Pumpsaint gold deposit in mid Wales occurring in both pyritic shales and quartz veins (Annels and Roberts, 1989). It also occurs in the Dolgellau gold belt in North Wales (Shepherd and

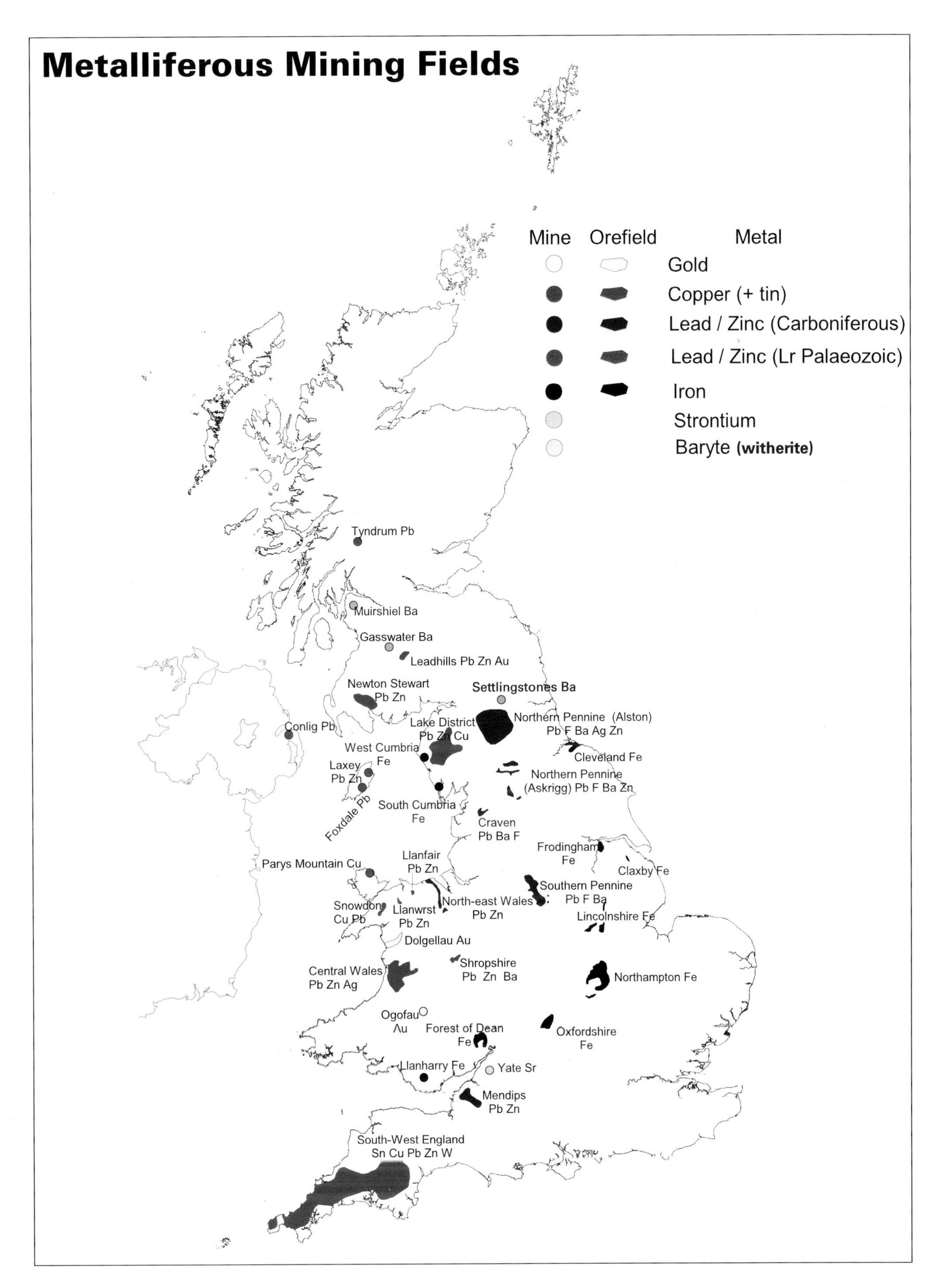

Figure 21 Old mining fields in Britain.

Bottrell, 1993). Small amounts of arsenopyrite were mined from Cambrian sandstones in the Tan-y-garth mine near Bethesda, North Wales (Dewey, 1920).

Scotland

The Glendinning antimony deposit in south-west Scotland was investigated by the BGS MRP in 1980 and four holes were drilled. These intersected stratabound pyrite-arsenopyrite mineralisation with up to 0.7% As over 12 m in vein wallrocks and in the matrix of intraformational breccias (MRP 59). Sporadic minor gold enrichment up to 840 ppb also occurred (Duller et al., 1997) which predates the antimony vein mineralisation described above. The deposit is considered to be similar to the Clontibret arsenic-antimony-gold deposit in Ireland and formed during the emplacement of late Silurian calc-alkaline minor intrusives at the close of the Caledonian orogeny. Arsenopyrite also occurs in the Caledonian intrusive-related copper mineralisation at Black Stockarton Moor, Moorbrock Hill, Fore Burn, Stobshiel and Glenhead Burn in the Southern Uplands, at Lagolochan near Oban and near Comrie (Pattrick and Polya, 1993). Additional information on the distribution of arsenic in the Southern Uplands is provided by Stone et al. (1995). Massive arsenopyrite lenses up to 75 cm thick, with associated gold, occur at Invergeldie, west of Comrie, at the contact of epidiorite sills and Dalradian metasediments (MEG 248). The Devonian hot-spring system near Rhynie in Aberdeenshire also contains enrichments in arsenic and antimony (Rice et al., 1995). Arsenic occurs in gersdorffite in the small Talnotry magmatic copper-nickel deposit in south-west Scotland (Stanley et al., 1987). Additional information on arsenic in Britain is given in Dewey (1920).

Barium

Baryte has been produced in several areas of Britain. The main deposits were:

1 Gasswater and Muirshiels in the Midland Valley of Scotland (Scott, 1967; MRP 67).
2 Settlingstones, a major producer of witherite, in Northumberland (Dunham, 1990).
3 In the Northern and Southern Pennine orefields (Dunham 1990; Dunham and Wilson 1985; Firman and Bagshaw, 1974).
4 In the Shropshire mining field (Dines, 1958).
5 More recently, the Foss and Edradynate deposits near Aberfeldy in Scotland (Beveridge et al., 1991).

Gasswater mine produced 0.5 Mt of baryte from fault- controlled veins in Devonian — Carboniferous sediments (Scott, 1967). Muirshiels mine produced 0.3 Mt from veins in Lower Carboniferous trachytic and basaltic lavas (MRP 67). Both mines are associated with a south-east trending swarm of Tertiary dolerite dykes but there appears to be no genetic connection. The mineralisation is thought to have originated from barium-rich formation waters or remobilisation of sedimentary baryte concentrations in Devonian–Carboniferous sequences (Gallagher et al., 1982). Vein deposits in Dinantian limestones and sandstones have been exploited in the Northern and Southern Pennine orefields for a total production exceeding 2 Mt. Small scale production is still maintained. The Settlingstones area in Northumberland was a major source of witherite ($BaCO_3$) with a total production of 740 000 t (Dunham, 1990). Several deposits were also worked in the Durham coalfield, to the east of the Northern Pennine Orefield, with a total production of 264 000 t of witherite and 234 000 t of baryte (Dunham, 1990). Substantial tonnages were also produced from the northern Lake District up to the 1960s , especially from the Ruthwaite mine (MRP 122). The Shropshire mining field yielded over 0.5 Mt of baryte from east–west veins up to 6 m wide in late Precambrian sediments (Dines, 1958). The baryte is accompanied by minor copper and bitumen minerals.

Figure 22 VULCAN 3D view of Foss baryte mine.

Modelling the MI(GB) Foss baryte orebody near Aberfeldy, Scotland using Maptek VULCAN software under the DTI Technology Access project

Recent production has centred on the Aberfeldy stratabound exhalative deposits (Coats et al., 1984b; MRP 26 and 40) which has large (several million tonnes) resources of high-grade baryte in Middle Dalradian sediments. Two separate deposits have been identified to the east and west of Ben Eagach. The smaller Foss deposit, to the west (Figure 22), is in production from an underground mine at the rate of around 50 000 t/yr. Proposals for mining the larger Duntanlich deposit to the east are awaiting planning approval. The mineralisation occurs near the top of the Ben Eagach Schist Formation, which includes graphitic schist, mica schist and quartzite with thin limestones. The mineralised zone is up to 110 m thick and extends over 7 km strike length. Bands of pure baryte over 15 m thick occur within the zone surrounded by an envelope of celsian (barium feldspar) and barian muscovite. Other promising prospects occur in the Dalradian to the north-east and south-west of Aberfeldy (Section 5.1.1). There has also been minor production from vein deposits in the Pennine orefields, in association with fluorspar extraction. and from vein deposits in Scotland at Strontian, Highland Region, and the Nethan Valley, near Lesmahagow, Strathclyde Region. Additional information on baryte deposits in Britain is given in Collins (1972).

Bismuth

South-west England

Minor amounts of bismuth have been recorded from south-west England (Dines 1956; Ball et al., 1982a; Embrey and Symes, 1987). However, only one mine has a recorded output of 1 ton of mixed bismuth, cobalt and nickel ore (Dines, 1956) and thus it appears that it was not present in any quantity in the region.

Lake District

Bismuth has been described from a number of quartz-sulphide veins in the Lake District (Stanley and Vaughan, 1982). It occurs in small amounts as native bismuth and bismuthinite in association with chalcopyrite-pyrite-arsenopyrite veins at numerous localities and with wolfamite-scheelite-arsenopyrite veins at Carrock Fell mine. A comprehensive list of bismuth localities in the Lake District is given in Young (1987).

Wales

Bismuth has been recorded in analyses of samples from the Rhos-mynach deposit in Anglesey (Dewey and Eastwood 1925). It is associated with chalcopyrite and minor gold in Silurian shales adjacent to a rhyolite intrusion. This small occurrence is about 4 km north-east of the Parys Mountain Zn-Cu-Pb VMS deposit and may be of similar origin. A lead-bismuth sulphide occurs in volcanogenic veins of Ordovician age at the Hafod-y-Porth mine in Snowdonia (Reedman et al., 1985).

Scotland

Bismuthinite occurs in late Caledonian granire-related mineralisation at Cairngarroch Bay (Lowry et al., 1997) and Foreburn (Charley et al., 1989) in the Southern Uplands; at Glen Gairn, Tomnadashan and Corrie Buie in the Dalradian terrane of western Scotland; and in Moinian rocks at Ratagain and Lairg in the northern Highlands (Pattrick and Polya, 1993). The Dalbeattie uranium veins, which cut Lower Palaeozoic sediments in the aureole of the Criffel granodiorite, contain minor amounts of bismuthinite (Miller and Taylor, 1966).

Chromium

Chromite has been produced from numerous small open-pit workings in the Unst Ophiolite Complex in northern Shetland. It occurs in pods up to 1 m thick and 75 m long in serpentinised dunite. About 50 000 t were produced up to 1945. More recent investigations by the BGS using a gravity survey to define drilling targets found no indications of significant near-surface chromite mineralisation (MRP 35). Chromite has also been recorded at Corrycharmaig in the Grampian Highlands (Harrison, 1985) and in north-east Scotland in small serpentinite bodies. It also occurs in the Ballantrae ophiolite (Stone et al., 1986) and in layered Tertiary ultrabasic rocks on Rum off the west coast of Scotland. Potentially large resources of chromite and olivine sand have been discovered by the MRP on the sea bed to the south of Rhum in marine deposits derived from the weathering and erosion of the adjacent Tertiary volcanic rocks (MRP 106).

Cobalt

Small amounts of cobalt have been recovered from south-west England in association with Sn-Cu mineralisation (Dines, 1956). Cobalt-bearing minerals such as siegenite ($(Co, Ni)_3S_4$) occur in the Central Wales Pb/Zn mining field in the 'early complex' vein assemblage of Mason (1997). It has also been reported from the Alderley Edge mines in Cheshire (Carlon, 1979), the Great Orme copper mines (Ixer and Stanley, 1996) and in the Lake District (Stanley and Vaughan, 1982).

Copper

Copper mineralisation occurs in many areas of Britain (Figure 23).

South-west England

This area was the world's main source of copper in the first half of the nineteenth century. The mineralisation occurs in veins spatially associated with Variscan granites. The main copper mineralisation generally occurred above the tin-bearing parts of the veins. Further information is included in section 5.9 and under tin in this appendix. A comprehensive account of mining in the region is given in Dines (1956).

Anglesey

The Parys Mountain Cu-Pb-Zn deposit, which has produced around 300 000 t of copper and was the world's largest copper mine in the early 1800s is described in Section 5.4.1. Further mineralisation in northern Anglesey is indicated by MRP investigations for gold and base metals (MRP 99).

Coed-y-Brenin and Glasdir

A porphyry-style copper deposit was discovered by RioFinex in 1968 at Coed-y-Brenin, about 5 km north of Dolgellau in North Wales, in Cambrian intrusive and sedimentary rocks (Rice and Sharp, 1976). A drill-indicated resource of about 200 Mt at 0.3% Cu has been determined. Shepherd and Allen (1985) provide additional information on the genesis of the deposit from fluid inclusion studies. Information on the occurrence and distribution of gold in the deposit in given in Miller (1993). The Glasdir deposit, about 2 km south of Coed-y-Brenin, is a mineralised breccia pipe in Cambrian sediments consisting of angular fragments of country rock veined and cemented by chalcopyrite, pyrite and quartz (Allen and Easterbrook, 1978; MRP 29). Total production was about 75 000 t at 1.5% Cu and minor gold. It may have formed at the same time as the porphyry-style deposit.

Dalradian pyrite belt

Stratabound pyrite mineralisation, with sporadic copper and copper-nickel enrichment in vein and stratiform deposits, occurs in thicknesses up to 200 m over a strike length exceeding 200 km (MRP 8 and 15). The mineralisation occurs within the Middle Dalradian Ben Lawers Schist Formation just above the Ben Eagach Schist Formation (Hall, 1993). Mineralisation in the Tyndrum (Ben Challum) and Knapdale (Meall Mhor) areas is described in Section 5.1.1.

Gairloch

Gairloch is a Besshi-type volcanogenic stratiform Cu-Zn deposit in Lower Proterozoic Lewisian metasedimentary and basic volcanic rocks (Jones et al., 1987; MEG 179). It was discovered by Consolidated Gold Fields Ltd in 1978 and has been intensively drilled, but is currently subeconomic. The MRP has investigated a number of adjacent similar prospects (MRP 146). Additional information is given in Section 5.4.2.

Vidlin

The Vidlin mineralisation is a stratabound volcanogenic massive sulphide Cu-Zn deposit in Lower Dalradian metavolcanic rocks. It was investigated by the BGS (MRP 4) and a mining company (MEG 150). No resource figure has been published but the estimated grade is 0.7% Cu and 0.5% Zn (Smith et al., 1984).

Kilmelford

The Kilmelford area contains porphyry-style mineralisation in Caledonian granodiorite. The mineralisation was discovered by Phelps Dodge Corporation and Noranda-Kerr Ltd in 1971 and investigated by the BGS in 1976 (MRP 9). No resource figures have been published but the mineralisation in the BGS holes is very low grade (0.03% Cu). BP Minerals International Ltd has completed an evaluation of the adjacent Lagolochan area as a gold target — see below (Harris et al., 1988; MEG 258).

Lake District

Chalcopyrite, with quartz, pyrite, arsenopyrite and, locally, some magnetite and Co-Ni-As sulphides, has been worked from several sites, most productively in the Coniston area. The mineralisation at Coniston occurs in breccia veins in the Ordovician Borrowdale Volcanic Formation. (Firman, 1978; Stanley and Vaughan, 1982; Millward et al., 1999).

Snowdonia

Minor Cu-Pb-Zn volcanogenic vein and stockwork deposits occur within the Snowdon caldera (Reedman et al., 1985). They are interpreted as filling volcano-tectonic faults but may also have been feeders to Kuroko-style massive sulphide deposits, though these are likely to have been removed by erosion. A Kuroko-style massive pyrite deposit at Cae Coch, north-east of Snowdon, may be the distal expression of an undiscovered buried proximal base-metal sulphide deposit (Ball and Bland, 1985).

Sediment-hosted red-bed copper deposits

Several small sediment-hosted red-bed copper deposits occur in the Cheshire Basin. The largest is the Alderley Edge deposit, which produced a recorded total of about 170 000 t of ore containing about 3200 t of recovered copper. The mineralisation is in Triassic sandstones and conglomerates and consists of baryte, chalcopyrite and galena impregnations and veins in areas of bleached or altered rock adjacent to faults (Warrington, 1966; Carlon, 1979). Small amounts of cobalt and vanadium have also been recovered. A comprehensive study of the entire Cheshire Basin, including its evolution, structural development, lithological provenance, hydrology and resource potential for base-metals, industrial minerals, hydrocarbons and groundwater has recently been completed by Plant et al. (2000).

Dinantian limestone

There are several small replacement copper deposits in Dinantian limestone at Middleton Tyas, near Ripon in northern England (MRP 54), at Ecton on the west side of the Southern Pennine Orefield (Robey and Porter, 1972) and at Llandudno on the North Wales coast (Williams, 1993). The Ecton deposit occurs within a 400 m long vertical pipe-like body of massive calcite and chalcopyrite (Critchley, 1979) and produced around 15000 t of copper in the eighteenth century. The Llandudno deposit occurs over a strike length of 100 m, with chalcopyrite mineralisation occuring as vein fillings in dolomitised limestone (Ixer and Davies, 1997).

Further information on copper deposits and occurrences can be found in Dewey and Eastwood (1925) and Allen (1980).

Fluorite

Economic fluorite mineralisation is restricted to the Dinantian limestone areas of Britain where it is associated with Pb-Ba-Zn mineralisation (Ixer and Vaughan, 1993). The two main producing areas have been:

Northern Pennine Orefield

The most important areas were around Weardale in the Alston area (Dunham, 1990) and around Greenhow in the Askrigg area (Dunham and Wilson, 1985). Mining finally ceased in Weardale in 1998. Fluorite occurs in oreshoots within major veins up to 20 km long and 10 m wide, and also in stratabound replacement orebodies (flats) adjacent to veins as well as in numerous minor veins (Greenwood and Smith, 1977). The Alston area shows pronounced zonation of some of the ore minerals. Fluorite dominates the inner zone with calcite and then baryte towards the margins of the mineralised area. This zonation is coincident with the subcrop of the concealed high-heat-production Caledonian Weardale granite. The Pennine mineralisation is considered by Dunham (1983) to be a fluoritic subtype of the Mississippi Valley Type (MVT) ore deposit.

Southern Pennine Orefield

The F-Ba-Pb mineralisation occurs in major east–west veins (rakes) and stratabound replacement deposits (flats) together with some cave fill deposits (pipes). The richest mineralisation is concentrated in the uppermost Dinantian limestones beneath Namurian shales on the east side of the Dinantian outcrop (Ford and Ineson, 1971; Ford, 1976). Current production is mainly from open-pit operations and is principally confined to the north-eastern part of the orefield between Eyam and Bradwell. This area is almost entirely within the Peak District National Park. Substantial (in excess of several hundred thousand tonnes) replacement orebodies have been worked, adjacent to major veins in the Bradwell area (Butcher and Hedges, 1987). They have been formed by the hydrothermal stoping of limestone causing collapse breccias which were subsequently replaced by fluorite (Jeffrey, 1997). They show similarities to MVT deposits in the USA as described by Briskey et al. (1986) and there is potential for further discoveries. There are also several small tributory operations feeding a central processing plant near Stoney Middleton. Smith (1988) reviews the planning history of the Milldam underground mine which opened in 1987 at Great Hucklow and additional information on mineral operations within the Peak District National Park is given in Anfield et al. (1998). A comprehensive survey of the minerals of the Southern Pennine Orefield is provided by Ford et al. (1993).

Fluorite also occurs, but in relatively small quantities, in some of the south-west England east–west Sn-Cu fissure veins and also the north–south Pb-Zn 'cross-course' veins (Alderton, 1993); as a cement in Permo-Triassic sandstones near Elgin, Grampian (Peacock et al., 1968); associated with the iron ore mineralisation in west Cumbria (Goldring and Greenwood, 1990) and in veins marginal to the Cairngorm granite.

Gemstones

Gemstones have not been produced commercially in Britain though several varieties of semi-precious stones, including

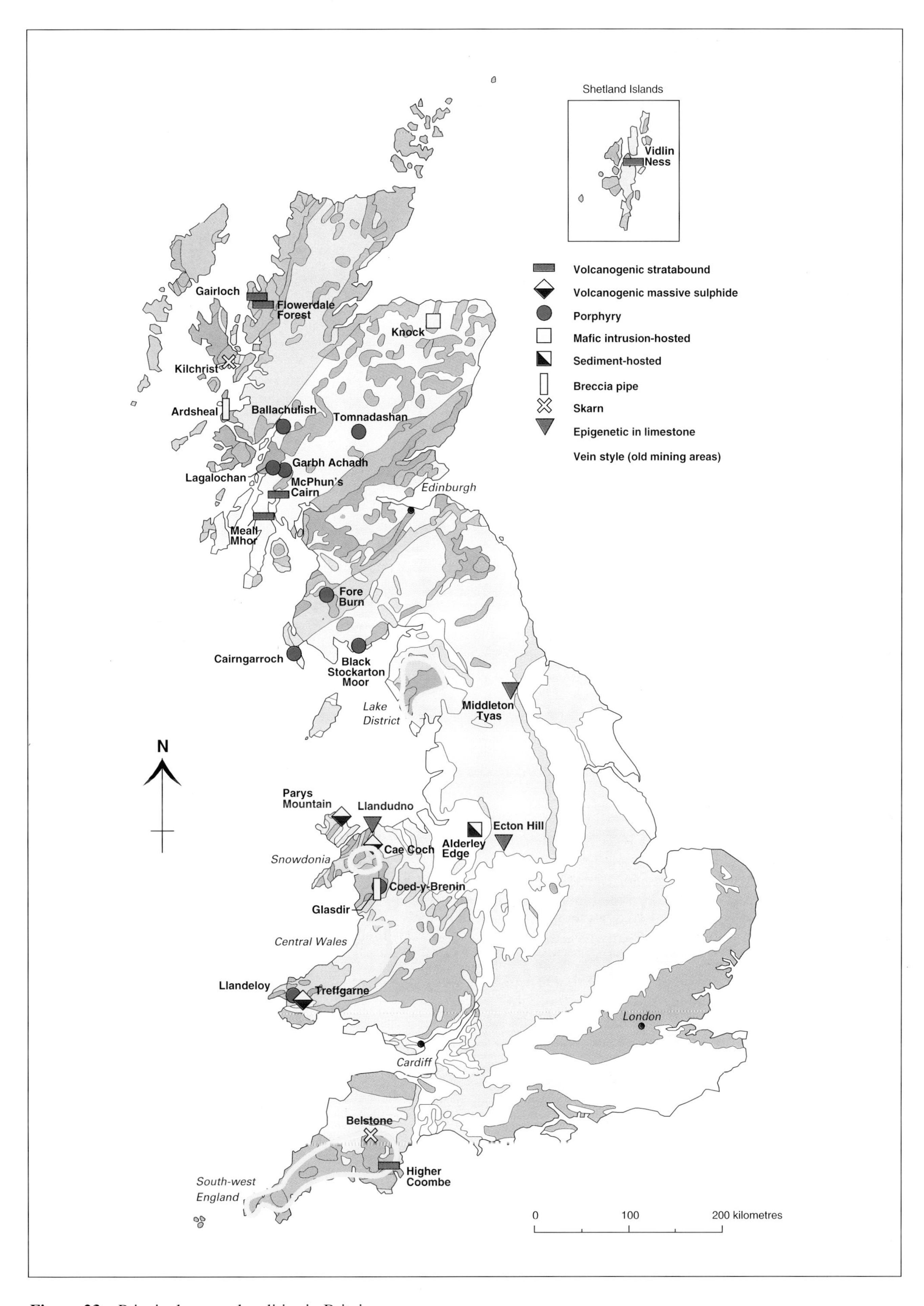

Figure 23 Principal copper localities in Britain.

quartz varieties (cairngorm and agate), garnet, fluorite and jet, have been worked in the past. Diamonds have not been reliably recorded from Britain but a recent MRP report suggested that some areas, such as the Lewisian terrane in north-west Scotland, may possess suitable tectonic conditions for the formation of diamonds (MRP 135). Sapphires have been known in Scotland for many years, but until recently were only found as small platy crystals from thermally altered aluminous sedimentary rocks associated with Tertiary basalts on Arran, Mull and Ardnamurchan and from Dalradian rocks in Aberdeenshire. However, in 1984, sapphire megacrysts 10–30 mm in size were discovered in a 1.5 m wide monchiquite dyke cutting Archean gneisses at Loch Roag on the Isle of Lewis in the Outer Hebrides. A 9.6 carat gem was cut from this material. Beryl has been recorded in micaceous pegmatites of the Moine schists in north-west Scotland (Berridge, 1969) with crystals up to 7 cm long at Struy Bridge. Topaz crystals weighing more than 0.5 kg have been found in the Cairngorms associated with quartz in granite pegmatites.

Gold

Several areas of Britain have produced minor amounts of gold (Figure 25). The most important have been the Dolgellau area of North Wales, where auriferous quartz veins in Cambrian shales have yielded a total of about 4 t of gold, and in Central Wales where auriferous quartz veins cut pyritic Silurian shales. Alluvial gold has been worked near Helmsdale in northern Scotland where it occurs overlying Precambrian Moine schist. Crummy et al. (1997) provide evidence for a possible epithermal source for this mineralisation, associated with Devonian volcanism. Gold has been panned from many streams in the Southern Uplands of Scotland, especially around Leadhills. Further historical information on gold in Britain can be found in Collins (1975).

North Wales

The Dolgellau Gold Belt occurs within Middle and Upper Cambrian sediments with numerous intrusions of 'greenstone' (basic to intermediate volcanic rocks) on the southern and eastern flanks of the Harlech Dome. The gold is associated with massive north-east trending mesothermal quartz veins up to several metres wide and several kilometres long (Shepherd and Bottrell, 1993). The distribution of gold is very sporadic. At Clogau St. David's mine it occurs in small, rich, steeply-dipping shoots associated with pyrite and pyrrhotite together with galena and bismuth tellurides. The veins are often auriferous where they cut graphitic shales of the Clogau Formation but not where the host rocks consist of non-graphitic arenites. Grades have been poorly recorded but averaged over 30 g/t Au in 1900 with very selective mining. Gold shoots generally occur at the intersections of side lodes, where greenstone and shale form the wall rocks and at braided structures with enclosed lenses of shale (Shepherd and Bottrell, 1993). The Clogau St. David's (MEG 244) and Gwynfynydd mines have recently been further evaluated Some development and production for jewellery has been pursued, but both mines are now closed. Grades from bulk samples of the Clogau St. David's evaluation range up to 7.9 g/t Au (Bottrell et al., 1988).

Central Wales

The Ogofau mine near Pumpsaint produced gold from quartz veins cutting pyritic Silurian shales close to the Ordovician–Silurian boundary. The shales are folded in an anticline and cleaved with development of crosscutting and saddle-reef quartz veins. Gold has been remobilised into some of the quartz veins and into fractures in pyrite and arsenopyrite (Annels and Roberts, 1989). Grades were around 7 g/t Au during the last period of working in 1938 but recovery was poor. Considerable research has been carried out since 1973 by the Department of Geology at Cardiff University, and the area is now under active exploration by Anglesey Mining.

Dalradian quartz veins

Gold bearing, discordant, steeply dipping quartz veins with minor sulphides occur within Middle Dalradian rocks in the Sperrin mountains of Northern Ireland (Clifford, 1986) and near Tyndrum in the Central Region of Scotland. Their age is unknown but at Tyndrum they appear to be related to a late Caledonian extensional phase, with plutonic activity, associated with the main east–west Tyndrum fault (Pattrick et al., 1988). They are earlier than the main north-east–south-west lead vein at Tyndrum which may be Carboniferous in age. Not all the quartz veins are auriferous and the Tyndrum main vein, previously worked for lead and zinc, does not carry significant gold. Individual veins are up to 100 m long and 2–3 m wide within zones up to one kilometre long and 100 m wide. Gold is distributed throughout mineralised quartz veins, not just in sporadic shoots. Three deposits, Curraghinalt (Earls et al., 1989) and Cavanacaw — formerly known as the Lack deposit (Cliff and Wolfenden, 1992) in Northern Ireland and Cononish in Scotland (Parker et al., 1989), are under active development to establish ore reserves. Geological ore reserves are reported to be 907 000 t at 10.3 g/t Au (cut) for Curraghinalt and 861 000 t at 7.54 g/t (cut) for Cononish. Cononish and Cavanacaw have both received planning permission for development and are awaiting suitable economic conditions to commence mine development. Base metal exploration in the Tyndrum area, including the Eas Annie lead mine, adjacent to Cononish, is described in MRP 50. Gold was also reported from minor quartz veins, with pyrite, chalcopyrite and galena, at Stronchullin in Knapdale, Strathclyde. Two 10 t samples contained 45 and 37 g/t Au, together with silver, arsenic and antimony (Peach et al., 1911).

Caledonian granites

Several Caledonian granite-granodiorite intrusions within Dalradian metasediments contain gold associated with sulphide-bearing fault and breccia zones with values up to 10 g/t Au (Rice, 1993). Host rocks are commonly strongly hydrothermally altered. They include Kilmelford (MRP 9), the neighbouring Lagalochan intrusion (Harris et al., 1988; MEG 258) in western Scotland and Comrie in central Scotland (MEG 226). A detailed follow-up stream-sediment and water survey of metal dispersion from the Lagalochan prospect is given by Polya et al. (1995). Further information is given in Section 5.2. There is a similarly mineralised intrusion in Lower Palaeozoic greywackes at Foreburn in the Southern Uplands adjacent to the Southern Uplands Fault (Charley et al., 1989 and MRP 55). The Hare Hill (Boast et al., 1990; MEG 135 and 257) and Moorbrock Hill (Naden and Caulfield, 1989; MEG 261) intrusions host north-south structures with quartz-antimony-arsenic mineralisation and some gold in earlier quartz veins.

Ochil Hills, Midland Valley of Scotland

Several alluvial gold localities were discovered by the MRP in the Ochil Hills during 1978. Subsequent detailed sampling and shallow drilling in the Borland Glen area identified intense hydrothermal alteration and brecciation indicative of a large epithermal system in Lower Devonian volcanics (MRP 116). Additional exploration, including limited production of alluvial gold, was subsequently carried out by a mining company.

Southern Uplands

Gold is associated with Lower Palaeozoic greywackes and shales at several localities in the Southern Uplands (MRP 28). At Glendinning mine stratabound sulphide (arsenopyrite) horizons host minor gold while cross-cutting quartz-antimony base-metal sulphide veins are non-auriferous (Duller et al., 1997; MRP 59). At Glenhead, adjacent to the Loch Doon granodiorite (MRP 48; MEG 261), and in the Leadhills area, quartz veins and disseminated arsenopyrite zones are auriferous (MEG 230 and 260). RioFinex examined Glengonnar Water, near Leadhills, for alluvial minerals, including gold (MEG 127).

Rhynie, Grampian

Anomalous levels of gold (up to 1 g/t Au), with silver, arsenic and antimony, occur in a silicified fault zone near Rhynie, Grampian (Rice et al., 1995) on the western margin of the Rhynie Devonian sandstone outlier. The zone is associated with conglomerates and chert (the 'Rhynie chert') which are interpreted as epithermal eruption breccias and volcanic sinter.

Tayside Region

Alluvial gold has been recorded in several streams in an area of Dalradian metasediments to the south of Loch Tay and from small base-metal workings in the vicinity (Heddle, 1901). These showings may be related to granite intrusions such as those at Comrie and Tomnadashan (Fortey, 1980) or quartz veins such as those at Corrie Buie where small inclusions of electrum occur in galena (Pattrick, 1984). Company exploration at Invergeldie, west of Comrie, located small outcrops of massive arsenopyrite up to 0.75 m thick at the contact of epidiorite sills and Dalradian metasediments (MEG 248). The arsenopyrite contained up to 19 g/t Au. However, drilling showed the sulphide was not extensive. MRP exploration in the Glen Clova area found a gold-bearing fault zone up to 1.6 km long in the Burn of Fleurs with up to 7 ppm Au in a clay fault gouge (MRP 126).

Central England

Traces of gold has been recorded in quartz veins cutting in late Precambrian diorites in Charnwood Forest, near Leicester, on the north-eastern margin of the Midlands Microcraton (King, 1968). A small gossan of cavernous quartz with limonite and traces of gold was found in the same area at Bardon Hill in 1950 (King, 1967). Analysis of alluvial gold grains from the Charnwood area indicates derivation from oxidising red-bed–type gold-copper mineralisation (MRP 144).

Devon

Dendritic gold occurs in calcite veins in Devonian limestones at Hope's Nose, near Torquay in south Devon. Exploration by the BGS has found numerous gold anomalies in drainage and overburden samples (Figure 24) from Lower Devonian sediments and intrusives in the South Hams area around Kingsbridge, 20 km south-west of Torquay (MRP 98). Some bedrock mineralisation has also been found. The gold in both areas is notably palladium-rich. Further exploration located additional anomalous areas in the Crediton Trough, north-west of Exeter, associated with Permo-Triassic red-bed sediments and Permian alkaline volcanics (MRP 133 and 134). Commercial exploration, including drilling has been carried out by Crediton Minerals Ltd. Lead and antimony vein mineralisation, with silver-rich galena and reported gold values, has been worked in the Wadebridge area of north Devon (Dines, 1956). The veins are in Devonian slates and volcanics. MRP investigations, including drilling, found gold at levels up to 1 g/t Au (MRP 103).

Investigations in the Sourton Tors area of north Dartmoor, over the Dinantian Meldon Chert Formation, located a zone of sulphide veins with pyrite and pyrrhotite but without significant base-metal mineralisation (Beer and Fenning, 1976). Two sulphide concentrates from drill core contained 5 and 40 g/t Au. A gold- and silver-bearing pegmatite dyke has been reported from the Dartmoor granite, on Wind Tor south of Widecombe (Brammall, 1926). Analyses showed up to 14 g/t Au and 23 g/t Ag.

Placer gold

In addition to the deposits in the Helmsdale, Ochils and Southern Uplands areas mentioned above, placer gold has been recovered as a by-product of alluvial tin mining in south-west England (Camm, 1995). No significant onshore placers are currently known, but some offshore areas have been investigated such as the Mawddach estuary, near Dolgellau in North Wales, which was considered a prospect by RioFinex during the late 1960s.

The BGS has carried out extensive analyses of alluvial gold grains throughout Britain and elsewhere using advanced microprobe techniques (Leake et al., 1993; 1995; 1997; 1998). Grains can be zoned (Figure 26) or contain characteristic inclusions of other elements, such as selenides, arsenides and PGEs, which vary with the provenance of the grain (Leake et al., 1995).

Figure 24 Gold Genie in operation in South Devon. Mechanised panning for gold from soil pits in the South Hams area.

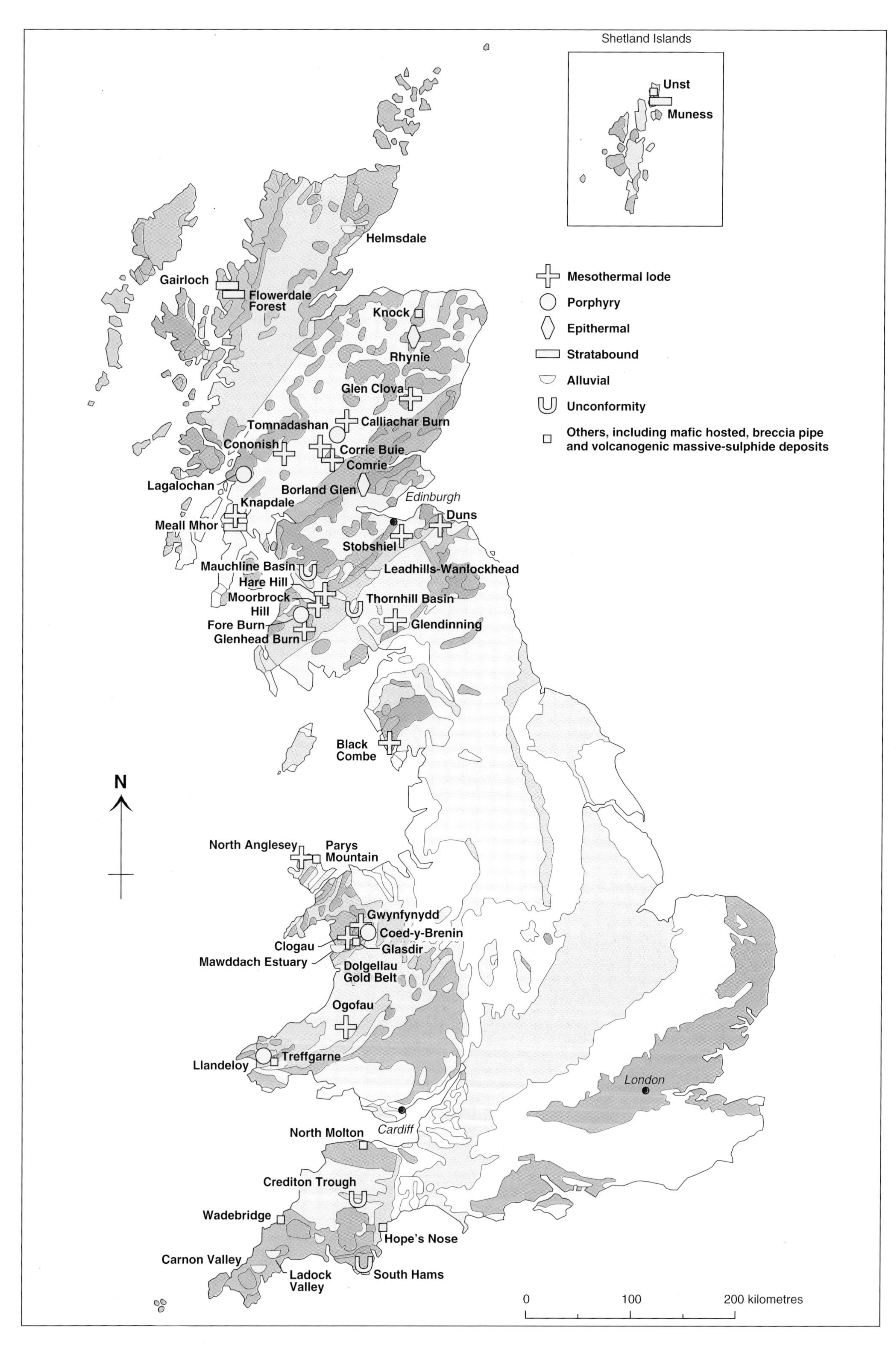

Figure 25 Principal gold localities in Britain.

Figure 26 Gold grain analysis. Microchemical maps showing several stages of growth, indicated by variable gold, silver and copper contents.

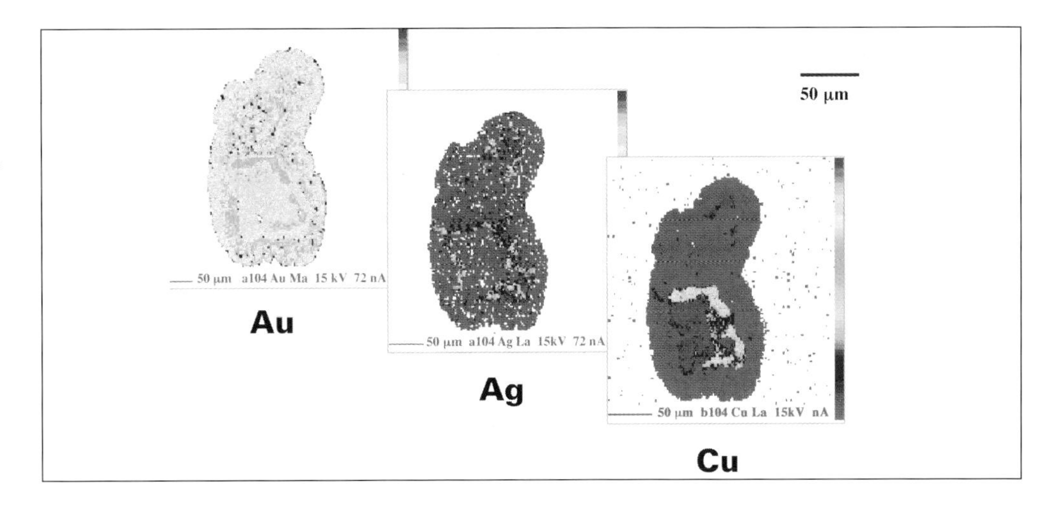

For example, gold grains associated with the Permo-Triassic red-bed style frequently contain palladium and also copper (MRP 144). The grains also contain inclusions of selenide minerals, and sometimes tellurides and arsenides, as these are stable under more oxidising conditions than the corresponding sulphide minerals which are generally absent from the Permo-Triassic gold grains.

Iron

Britain has been a major producer of iron ore, but only small amounts are mined now. The main types of ore worked and areas of production were:-

1. Replacement hematite deposits in Carboniferous Limestone in west and south Cumbria (Rose and Dunham, 1977; Shepherd and Goldring, 1993), in South Wales (Gayer and Criddle, 1969) and the Forest of Dean (Sibly, 1927). Total production has been about 250 Mt from the Cumbria orefields, 8 Mt from South Wales and 10 Mt from the Forest of Dean (Dunham et al., 1979). The Florence mine, near Egremont in Cumbria, is the only remaining mine and produces around 1000 t of hematite per year for pigments.
2. Jurassic bedded sedimentary low-grade (20 to 35% Fe) ironstone deposits in the English Midlands (Young, 1993). These were worked on a large scale at the rate of 10 to 15 Mt/yr up to 1970 (Dunham et al., 1979). Production has now ceased.

There was also substantial (for the time) production from Westphalian Coal Measures 'Blackband' and 'Clayband' ironstones, especially in Staffordshire and the Midland Valley of Scotland (Strahan et al., 1920) and from vein and replacement deposits in the Northern Pennine Orefield (Dunham, 1990). Ordovician oolitic bedded iron ores were worked in North Wales (Trythall et al., 1987). Banded quartz-magnetite lenses occur in Lewisian (Archean to early Proterozoic) rocks near Gairloch and on the island of Tiree (MRP 146). Small iron skarn deposits occur at Clothister Hill on Shetland and on the Isle of Skye with associated copper mineralisation (Groves, 1952). A summary of British iron ore deposits is given in Slater and Highley (1976).

Lead and zinc

Lead and zinc mineralisation occurs in many areas of Britain (Figure 28). The main types of deposit are:

Dinantian carbonate-hosted deposits

Mineralisation generally occurs in veins and replacement deposits, similar to those described under fluorite, which commonly accompanies the sulphide mineralisation in many of the orefields. The main producing areas have been the Northern (Dunham, 1990; Dunham and Wilson, 1985) and Southern (Ford, 1976) Pennine orefields, North Wales (Earp, 1958) and the Mendips (Green, 1958; Ford, 1976). A recent overview of all these areas is provided by Ixer and Vaughan (1993). They were mainly worked for lead and the Pennine orefields were later worked for baryte and fluorite. Zinc minerals occur but were often discarded or not recorded. There was some zinc production from the Nenthead area of the Northern Pennine Orefield (Dunham, 1990) and from North Wales (Earp, 1958). The silver content of the galena is generally low but ranges from 10 g/t in the Southern Pennines up to 550 g/t in North Wales. Mineralisation is thought to have commenced in early Permian times and extended through to the Triassic or early Jurassic in some areas. The source of the mineralising fluids is generally thought to be Carboniferous shale basins adjacent to the carbonate platforms which host the mineralisation. Fluid movement occurred during the Variscan orogeny aided by seismic pumping from the over-pressured basins along listric faults (Plant et al., 1988). Minor Pb-F mineralisation occurs in Dinantian carbonates intersected by oil exploration boreholes in eastern England (Ineson and Ford, 1982) indicating the possible presence of concealed mineralisation to the east of the Pennine orefield. Pb-Zn-Ba mineralisation in Dinantian limestone has been interesected in the Bruton drillhole in the Wessex Basin (Holloway, 1982). Stratabound, very low grade lead and zinc mineralisation occurs in basal Dinantian limestones on the northern margin of the Northumberland basin (MRP 17; Gallagher et al., 1986). No veins are currently worked for lead and zinc alone but minor amounts of lead and zinc are produced as a byproduct of fluorite and baryte mining.

Lower Palaeozoic greywacke and volcanic-hosted deposits

The main producing areas have been Central Wales, north-west Wales (Llanwrst), the Lake District, Shropshire (Dines, 1958), the Isle of Man (Lamplugh, 1903), Tyndrum in the Southern Highlands of Scotland (Curtis et al., 1993) and the Southern Uplands of Scotland (Leadhills). One of the largest lead mines in Britain, the Greenside mine in the Lake District (Gough, 1965), produced over 200 000 t of lead concentrates before closure in 1962. Mineralisation shows a spatial association with underlying granites in the Lake District and the Isle of Man, and with volcanics in

Figure 27 Sphalerite (brown) in vein breccia from Central Wales.

North Wales (Haggerty, 1995), but in Central Wales and most of the Southern Uplands there is no indication of magmatic involvement.

Phillips (1986) has described how metamorphic fluids could have given rise to the vein mineralisation in Central Wales. The Central Wales mineralisation occurred in two main episodes at 390 Ma and 360 Ma (Fletcher et al., 1993). The earlier mineralisation contains a complex, fine-grained series of Pb-Zn-Co-Ni sulphides, while the later episode (Figure 27) is composed of open space filling, coarse-grained, brecciated, simple Pb-Zn sulphides (Mason, 1997). A thorough review of the Central Wales mineralisation is given by Jones (1922) and more recent mineralogical and textural information by Raybold (1974) and Mason (1997). The galena in the Lower Palaeozoic deposits generally contains more silver than that in the carbonate-hosted mineralisation and the associated gangue minerals are generally quartz and/or calcite.

Stratabound Dalradian deposits

Sphalerite and galena have been found in stratabound lenses in the Middle Dalradian Ben Eagach Schist Formation at Aberfeldy in the Tayside Region of Scotland (MRP 40; Coats et al., 1980). The highest reported grades are 8.5% Zn and 3.6% Pb over 4.3 m true width. There are a number of other occurrences of stratabound lead and zinc sulphides in Middle Dalradian metasediments (MRP 88, 93 and 104; Coats et al., 1984b).

Mississippi Valley-type mineralisation in the North Sea.

Offshore hydrocarbon exploration drillholes in the Moray Firth Basin of north-east Scotland (Figure 28) have proved Pb-Zn-Ba mineralisation in Jurassic sandstone reservoirs (Baines et al., 1991). The mineralisation occurrs within the oil zone of the reservoirs and can form up to 30% of the toal rock volume as late cements.

Fissure veins in south-west England

Galena and sphalerite have been worked from the north-south 'cross-course' fissure veins in south-west England especially in the Liskeard and Callington areas (Dines, 1956). They usually occur more than 1 km from the major granite intrusions and are generally deposited from lower temperature fluids (Alderton, 1993). The galena was often silver-rich averaging 40 oz per ton (~1200 g/t). Total production was around 300 000 t of 'lead ore' containing 60–75% Pb (Dines, 1956). Zinc was produced as a co-product of tin production at Wheal Jane mine in Cornwall until it closed in 1991.

North-west Scotland

Noranda-Kerr investigated the potential of the Cambrian Durness Group quartzites and dolomitic limestones for stratabound lead and zinc deposits of Laisvall type (Scott, 1976; MEG 85 and 113) along the line of the Moine Thrust. Weak disseminated galena mineralisation was found in two areas near Loch Ailsh (MEG 113), and two boreholes were drilled south-west of Achnashellach station through the basal quartzites, but no significant mineralisation was found (MEG 85).

Variscan Front

Other lead and zinc localities are associated with the Variscan Front in southern England and South Wales (Figure 28). Apart from the well-mineralised Mendip Hills (Green, 1958), minor Pb-Zn-Ba mineralisation occurs in drillholes in the Wessex Basin (Holloway and Chadwick, 1984) and at a number of localities in South Wales (Fletcher, 1988). Although hosted mainly in Dinantian carbonates, the age of mineralisation spreads from Triassic into Lower or Middle Jurassic times. In the Mendip Hills many deposits were worked in the Triassic Dolomitic Conglomerate and in South Wales narrow veins extend into the basal Jurassic sediments (Fletcher, 1988)

Lithium

South-west England granites have a high lithium content, mainly contained in lithium-bearing micas. In particular, parts of the St Austell granite form a large, but low-grade, lithium resource containing micas with up to 2.5% Li (Hawkes et al., 1987). It is estimated that about 3 Mt of lithium at a grade of 0.15% Li are contained within an 8 km^2 area to a depth of 100 m. The St Austell granite has been extensively worked for china clay and production continues.

Manganese

Manganese has been produced in two areas in North Wales. An extensive 40 cm thick bed of syn–diagenetic origin has been worked in Cambrian pelites around the Harlech Dome (Bennett, 1987). Deposits of probable volcanic-exhalative origin occur in Ordovician mudstones and dolerites at Rhiw in the Lleyn Peninsula. The MRP investigated this area using magnetic surveys and drilling but only minor mineralisation was found (MRP 102). Total production from North Wales is about 135 000 t (Rhiw) and 44 000 t (Harlech).

There are numerous minor localities in Scotland. The largest is the Lecht deposit, near Tomintoul in Grampian. It is a manganiferous ironstone breccia in Dalradian metasediments and has been interpreted as a gossan overlying stratiform sulphides (Smith, 1985). However, it is now considered to be an ironstone derived from the weathering of an exhalative, spessartine-bearing protore (Nicholson, 1987; Nicholson and Anderton, 1989). Manganese nodules with over 10% Mn occur in Loch Fyne on the west coast of Scotland but resources are probably small.

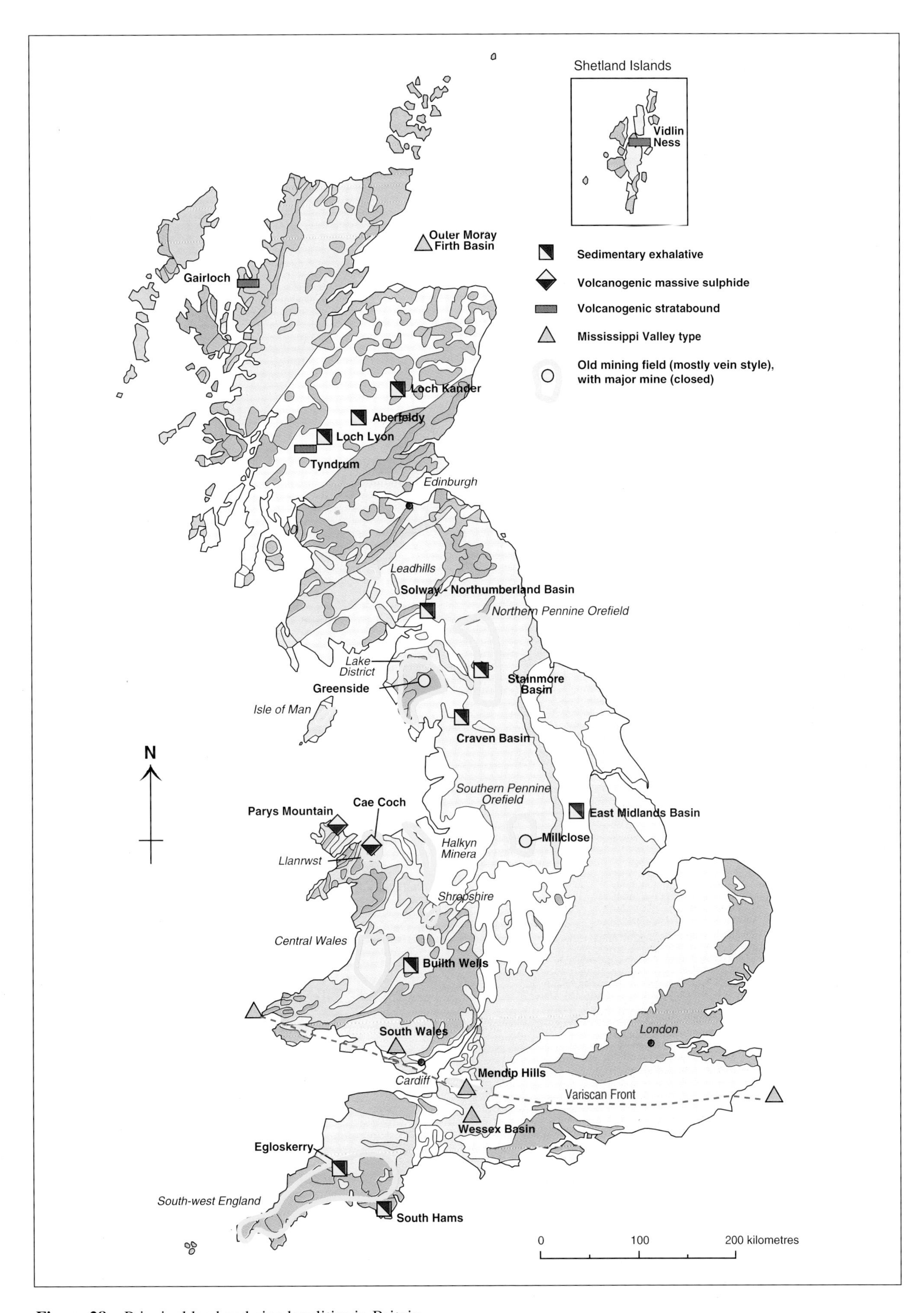

Figure 28 Principal lead and zinc localities in Britain.

Additional information on manganese localities in Britain is given in Woodland (1956) and Groves (1952).

Molybdenum

Molybdenum has not been worked commercially in Britain. Numerous minor occurrences are known, generally in veins and disseminations associated with Caledonian acid magmatism. MRP 3, 9, 43, 76 and 100 describe some Scottish localities. They include the Lairg, Kilmelford, Ballachulish, Glen Etive and Inverurie (Chapel of Garioch) areas. Company exploration data are on open file in MEG 3 and 17. Molybdenum also occurs in the Shap Granite in the Lake District (Firman, 1978; MRP 109) and in south-west England (Dines, 1956).

Nickel

A major exploration programme was carried out between 1965 and 1973 by Exploration Ventures Ltd (RioFinex and Consolidated Gold Fields) over the large basic-ultrabasic intrusions of Grampian Region in north-east Scotland (Figure 29). The initial exploratory work is described by Rice (1975). Cu-Ni sulphide mineralisation was discovered near Huntly (Littlemill zone) and Ellon (Arthrath zone) in Caledonian norites contaminated by pyritic and graphitic Dalradian sediments. The relatively low grade (less than 2% Cu+Ni) and difficulty of correlating drill intersections caused the project to be abandoned. However, substantial drill-indicated resources were located, with 2 Mt at 0.52% Ni and 0.27% Cu at Littlemill and 17 Mt at 0.21% Ni and 0.14% Cu at Arthrath (Fletcher et al., 1997). There are numerous data in the MEG Open File projects (Section 5.6). Exploration has also been carried out on the ophiolite complexes (Section 5.6) of the Lizard, Ballantrae and Shetland where high-tenor PGE enrichment has been discovered. Drilling at Ballantrae by Selection Trust Ltd found disseminated and massive nickeliferous marcasite over widths of a few metres (MEG 103).

Nickel-copper mineralisation at Talnotry, about 8.5 km north-east of Newton Stewart in south-west Scotland, was mined on a trial basis between 1885 and 1900. About 100 t of ore was raised, but not taken away (Gregory, 1928). The mineralisation, thought to be of magmatic origin, occurs at the base of an altered diorite sheet close to its contact with hornfelsed Lower Palaeozoic greywackes (MRP 10). The principal ore minerals are nickeline, gersdorffite, pentlandite and chalcopyrite (Stanley et al., 1987) which are associated with local minor enrichments in Pt, Pd and Au. Geophysical surveys conducted by the BGS indicate that the mineralisation is restricted to a single lenticular lode about 20 m long.

Two former small copper mines near Loch Fyne in the Inverary area of western Scotland also produced small quantities of nickel from stratabound Fe-Cu-Ni sulphides in Dalradian metasedimentary and volcanic rocks (Hall, 1993). Over 400 t of nickeliferous mineralisation, containing pyrrhotite, chalcopyrite, pyrite and pentlandite were produced between 1854 and 1867 from the Coille-bhraghad deposit. Similar mineralisation occurs on the dumps of the Craignure mine, 10 km to the south-west (Wilson and Flett, 1921). The ore appears to be a replacement of quartzite bands within the predominantly pelitic Dalradian metasedimentary sequence. The extent of the mineralisation is unknown, but as the deposits occur at similar structural levels, both in proximity to an extensive suite of metabasic intrusives, potential may exist for further occurrences along strike between them. Several companies have examined this area without success (MEG 2, 4, 115 and 123).

Niobium (columbium) and Tantalum

These metals have not been produced in Britain. Minor amounts of columbite have been recorded in the Cairngorm granite in Grampian, Scotland (MRP 96) and in a pegmatite on South Harris in the Hebrides (Berridge, 1969). Niobium-tantalum minerals occur in the Meldon Aplite, north of the Dartmoor granite in south-west England (Knorring and Condliffe, 1984).

Platinum Group Elements (PGE)

PGE have been found in the Shetland ophiolite complex on Unst (MRP 73; Lord and Prichard, 1997) and in the basic intrusions of north-east Scotland (Section 5.6). Minor enrichment in PGE (to 210 ppb) was noted during MRP investigations of the South Harris anorthosite-gabbro complex (MRP Data Release No. 10). Platinum-group minerals have also been reported from the Tertiary layered igneous complex on Rum in western Scotland (Butcher et al., 1999). There is little data on PGE values from Ballantrae in south-west Scotland; the Lizard in south-west England or other ophiolitic rocks and layered intrusions that are known in Britain.

Rare earth elements (REE)

These have not been produced in Britain. The Loch Borralan alkaline igneous complex in north-west Scotland contains apatite-rich pyroxenites with elevated contents of REE (Notholt et al., 1985). Subsequent investigations by the MRP over the adjacent Loch Ailsh intrusion showed a maximum of 0.7% La+Ce in syenite (MRP Data Release No. 11). The same investigation also found a maximum of 2.5% La+Ce in syenite from the Cnoc nan Cuilean intrusion near Ben Loyal in northern Scotland. Panned concentrates from some stream sediments in Central Wales contain more than 1% total REE in the form of recrystallised detrital monazite nodules (Read et al., 1987 and MRP 130). Hyslop et al. (1999) describe a suite of REE-bearing minerals from the Tertiary granite on Arran in western Scotland.

Silver

More than 400 tonnes of silver has been produced in the Northern Pennine orefield from argentiferous galena containing 150 g/t Ag (Dunham, 1990). The Central Wales orefield, where the galena contained over 200 g/t Ag, has produced over 100 t (Jones, 1922). The Foxdale and Laxey mines on the Isle of Man were also notably silver-rich with the silver content of lead ore ranging from 9 to 70 oz (270 to 2100 g) per ton (Lamplugh, 1903) Small amounts of native silver were produced from veins cutting Devonian volcanics and Dinantian sediments in Central Scotland, at Alva and Hilderston respectively; these areas have been investigated by the BGS (MRP 53 and 68). The history of the Alva mine, and information on mineral samples, is given by Moreton (1996). Argentiferous galena was also worked in Devonian sediments at Combe Martin on the north Devon coast (MRP 90) and in the south-west England mining field from Pb-Zn 'cross-course' veins emplaced at lower temperature than, and often crosscutting, the earlier Sn-Cu veins (Dines, 1956). The mines with the highest silver content (sometimes exceeding 2000 g/t of lead metal) were in the Liskeard area of east Cornwall (Ridgway, 1983). Silver mineralisation, with associated galena, was worked in a small mine on Sark in the Channel

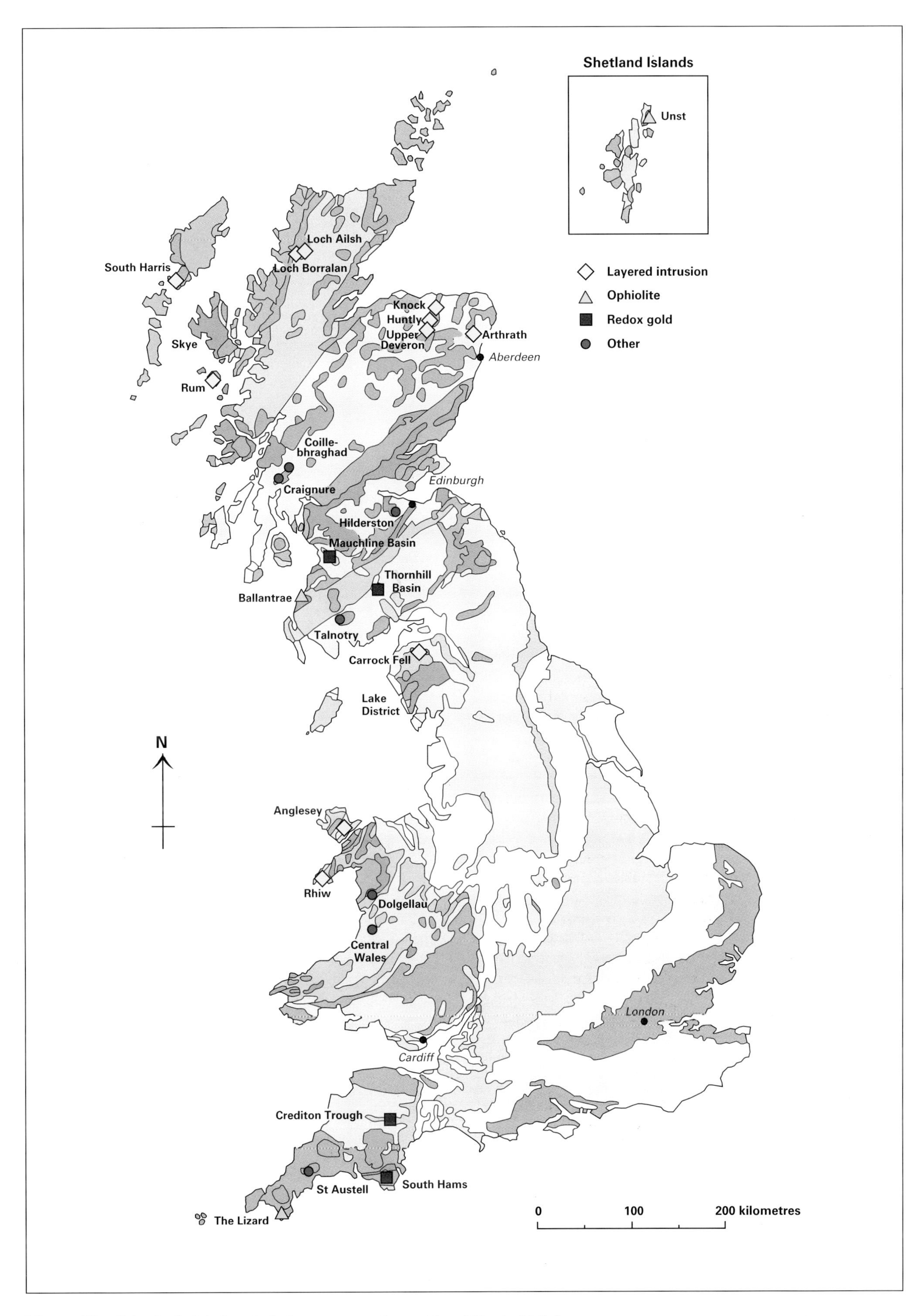

Figure 29 Principal nickel and platinum group element localities in Britain.

Islands (Ixer and Stanley, 1983). Further information on silver in Britain is given in Ridgway (1983).

Strontium

Strontium was first discovered at the type locality, Strontian, in western Scotland in strontianite associated with baryte. It was commercially extracted until the mid 1990s by open-pit in the form of celestine nodules from the Triassic Mercia Mudstone Formation near Bristol (Nickless et al., 1976). Total production of celestite exceeded 750 000 t making this area one of the largest sources of the element in the world. Further information on celestite in Britain is given in Thomas (1973).

Tin

Tin production in Britain ceased in 1998 with the closure of the South Crofty mine in south-west England. Over 2.5 Mt of tin-in-concentrates has been produced from veins associated with Variscan granites which intrude Devonian metasediments — locally known as 'killas'. Copper, zinc, arsenic and tungsten are associated with tin in quartz-tourmaline veins. There are several different styles of mineralisation including skarns, pegmatites, greisen veins, replacement deposits in calc-silicates, vein swarms, stockworks and pipes as well as the most important and ubiquitous fissure vein deposits. A comprehensive account of this classic mining region is given in Dines (1956). Later references include Dunham et al. (1979) and Anon. (1982). Embrey and Symes (1987) describe the mining history and minerals of this famous region while Alderton (1993) provides a modern synthesis of the metallogenesis. Recently operating mines were South Crofty, which worked a fissure vein swarm, and Wheal Jane, which worked an unusual deposit of cassiterite in very fine-grained sulphides, with sphalerite and minor chalcopyrite. The Wheal Jane deposit occured at the sheared contact of a shallow dipping felsite dyke 'elvan' and killas (Devonian metasediments). There are still major prospects in the region which are discussed in Section 5.9. Further examples of prospective areas can be found in Slater (1974). There is still controversy as to the ultimate origin of the tin and tungsten but Beer and Ball (1986) have shown that they are probably not derived from the sediments surrounding the granites. A number of papers relevant to mineralisation in the area can be found in Evans (1982) and Institution of Mining and Metallurgy (1985). An offshore spectrometric survey was carried out over the seaward extension of Geevor mine to detect additional mineralised lodes and to map the granite/killas contact (MEG 264).

Tin placers have been worked in many areas in south-west England. Some are associated with fossil strand lines at various levels up to 130 m above present sea level caused by subsidence during Pliocene times and subsequent uplift (Dunlop and Meyer, 1973). The area was not glaciated, but periglacial effects, combined with recent weathering and erosion, have formed tin placer deposits in many valleys. The geomorphic setting and history of the development of the stanniferous placers of south-west England is discussed by Camm and Hosking (1985) who also describe the styles and mineralogy of the primary tin mineralisation of the area. Most of the placers are worked out and the most prospective areas are offshore. Recent exploration projects are described in Section 5.11.3.

Small amounts of cassiterite occur in thin magnetite bands in the Carn Chuinneag igneous complex in northern Scotland (Gallagher et al., 1971). Anomalous levels of tin and tungsten (up to 1000 ppm) occur in thin quartz-magnetite veins in North Wales (Reedman et al., 1985) and traces of tin are associated with the Lake District granites (MRP 128).

Titanium

No titanium minerals have been produced in Britain. Potential hardrock sources, associated with basic igneous rocks, do not have sufficiently high concentrations to be of commercial interest at present. The main areas where titanium levels, mainly in ilmenite or titanomagnetite, exceed normal levels in basic rocks are:

North-east Scotland

The large Caledonian basic intrusions of Insch, Cabrach and Boganclough are enriched in iron and titanium, containing up to 5% TiO_2, mainly in ilmenite. No primary layered concentrations of Fe-Ti oxides have been found, although silicate layering is present in the ultrabasic parts of some of the intrusions. The area has been affected by deep Tertiary weathering and substantial amounts of ilmenite have entered the secondary environment, especially through fluvioglacial erosion. The effects of this can be seen in the titanium image in the G-BASE East Grampians atlas, where enhanced titanium values extend to the east of the basic intrusions. Heavy minerals, chiefly ilmenite and magnetite, are widely distributed in sands and gravels of the Grampian region especially in the Don Valley. Thin millimetre-scale heavy-mineral bands with ilmenite occur in coastal dune sands up to 10 m high near Rattray Head, between Peterhead and Fraserburgh (Colman, 1982).

Lake District

The small Caledonian Carrock Fell gabbroic intrusion contains ilmenite- and titanomagnetite-enriched zones in part of an east–west-trending complex close to the contact with an earlier acid granophyre body. Ilmenite and magnetite can form up to 20% of the rock, and chromite has also been recorded (Eastwood et al., 1968).

Other areas

Minor amounts of rutile have been recorded from beach sands in Northumberland (Gallagher, 1974). Glacial tills and residual soils in parts of the Lizard peninsula in Cornwall contain elevated levels of TiO_2, as ilmenite, over thicknesses of several metres. Drilling by the MRP has also revealed an extensive oxide-rich gabbro near Trelan on the Lizard (MRP 117) with up to 15% oxide (mainly as ilmenite).

Tungsten

Small amounts of tungsten have been produced in south-west England from early, high temperature quartz-tourmaline and quartz-feldspar veins. The four main producers were South Crofty, East Pool (wolframite with tin and copper), Hemerdon (wolframite with minor cassiterite) and Castle an Dinas (wolframite) (MRP 83). Major unworked low-grade stockwork deposits occur at Hemerdon (Section 5.9), which represents the largest resource of tungsten in Britain, and at Cligga Head on the north Cornwall coast. The origin of the tungsten mineralisation is discussed by Beer and Ball (1987).

The Carrock Fell mine in Cumbria was an intermittent, small- scale producter between 1901 and the early 1980s. It lies on the northern margin of the Skiddaw granite in a greisen zone with veins of quartz carrying wolframite, scheelite and sulphides (MRP 7, 33 and 60; Ball et al., 1985a). The last recorded head grade was 1.29% WO_3 (Slater, 1973). Wolframite and scheelite have been found in panned concentrates from the southern Lake District though the source has not been defined (MRP 128). Wolframite and scheelite also occur in trace quantities associated with molybdenite mineralisation around a small Caledonian granite near Inverurie in north-east Scotland (MRP 100) and in the Etive complex in western Scotland (MRP 76).

Uranium

Uranium occurrences are widespread in south-west England and Scotland. In south-west England, lower temperature pitchblende-hydrocarbon-sulphide mineralisation occurs within the main higher temperature Sn-Cu veins and in cross-courses associated with Pb-Zn-Co-Ni mineralisation. Uranium was produced as a minor by-product from a few copper and tin mines and one mine, South Terras, 5 km west of St Austell, was worked principally for uranium. Total production was only a few hundred tonnes of low-grade 'ore'. Ball et al. (1982b) give a comprehensive summary of the deposits.

In Scotland the most important uranium mineralisation occurs in low-grade, uraniferous, phosphatic and carbonaceous horizons in the Middle Devonian Orcadian lacustrine basin of the Orkneys and Caithness (Michie and Cooper, 1979), in Devonian arkosic breccias marginal to the Caledonian Helmsdale granite at Ousdale on the east coast of Caithness and in veins marginal to the Caledonian Criffel granodiorite at Dalbeattie in southern Scotland (Miller and Taylor, 1966). The Ousdale area was drilled on a 130 m square grid with 41 percussion holes to depths of 80 m (MEG 82). The maximum value found was 850 g/t U within a 15 m intersection. U-Pb mineralisation occurs in a fault breccia in Devonian sediments at Mill of Cairston near Stromness on Orkney (McCready and Parnell, 1998). The fault was drilled by the BGS and a mining company consortium in 1971–72 when maximum values of 0.1% U and 5.5% Pb were found (MEG 125). There are numerous other minor occurrences. Comprehensive summaries of uranium and other mineralisation in Scotland are given by Gallagher et al. (1971) and Michie and Gallagher (1979). Additional information on uranium occurrences in Britain is given in Appendix 4.

Vanadium

Minor amounts of vanadium were produced from the Alderley Edge copper mine in Cheshire where the element occurs in vanadinite (Carlon, 1979; Braithwaite, 1994). Pyroxenites within the Loch Borralan igneous complex in north-west Scotland contain 10% magnetite with up to 0.5% V_2O_5 (Notholt et al., 1985). Titaniferous magnetites from the Insch gabbro, one of a series of major basic intrusions in north-east Scotland, contain up to 1.7% V_2O_5. Ilmenites from the same rock contain up to 1% V_2O_5. However, the oxide mineral content of the rocks is generally low, at 5–10%. Permian mudstones in south Devon contain radioactive vanadiferous nodules with up to 14% V_2O_5. Mineralisation within the nodules consists of uranium, vanadium and iron oxides with base-metal sulphides. The results of a cored drilling programme are given in MRP 89. The gabbroic rocks of the Lizard were investigated by the MRP and drilling near Trelan proved the existance of an extensive oxide-rich gabbro containing 10 to 15% combined ilmenite and magnetite. Ilmenite is the main oxide and contains up to 5% V_2O_5 (MRP 117).

Appendix 2 Mineral Reconnaissance Programme Reports and Data Releases

Mineral Reconnaissance Programme Report Title

1 The concealed granite roof in south-west Cornwall.
2 Geochemical and geophysical investigations around Garras Mine, near Truro, Cornwall.
3 Molybdenite mineralisation in Precambrian rocks near Lairg, Scotland.
4 Investigation of copper mineralisation at Vidlin, Shetland.
5 Preliminary mineral reconnaissance of Central Wales.
6 Report on geophysical surveys at Struy, Invernesshire.
7 Investigation of tungsten and other mineralisation associated with the Skiddaw Granite near Carrock Mine, Cumbria.
8 Investigation of stratiform sulphide mineralisation in parts of central Perthshire.
9 Investigation of disseminated copper mineralisation near Kilmelford, Argyllshire, Scotland.
10 Geophysical surveys around Talnotry Mine, Kirkcudbrightshire, Scotland.
11 A study of the space form of the Cornubian granite batholith and its application to detailed gravity surveys in Cornwall.
12 Mineral investigations in the Teign Valley, Devon. Part 1–Barytes.
13 Investigation of stratiform sulphide mineralisation at McPhun's Cairn, Argyllshire.
14 Mineral investigations at Woodhall and Longlands in north Cumbria.
15 Investigation of stratiform sulphide mineralisation at Meall Mor, South Knapdale, Argyll.
16 Report on geophysical and geological surveys at Blackmount, Argyllshire.
17 Lead, zinc and copper mineralisation in basal Carboniferous rocks at Westwater, south Scotland.
18 A mineral reconnaissance survey of the Doon–Glenkens area, south-west Scotland.
19 A reconnaissance geochemical drainage survey of the Criffel–Dalbeattie granodiorite complex and its environs.
20 Geophysical field techniques for mineral exploration.
21 A geochemical drainage survey of the Fleet granitic complex and its environs.
22 Geochemical and geophysical investigations north-west of Llanrwst, North Wales.
23 Disseminated sulphide mineralisation at Garbh Achadh, Argyllshire, Scotland.
24 Geophysical investigations along parts of the Dent and Augill Faults.
25 Mineral investigations near Bodmin, Cornwall. Part 1–air borne and ground geophysical surveys.
26 Stratabound barium-zinc mineralisation in Dalradian schist near Aberfeldy, Scotland: preliminary report.
27 Airborne geophysical survey of part of Anglesey, North Wales.
28 A mineral reconnaissance survey of the Abington– Biggar–Moffat area, south-central Scotland.
29 Mineral exploration in the Harlech Dome, North Wales.
30 Porphyry-style copper mineralisation at Black Stockarton Moor, south-west Scotland.
31 Geophysical investigations in the Closehouse–Lunedale area.
32 Investigations at Polyphant, near Launceston, Cornwall.
33 Mineral investigations at Carrock Fell, Cumbria. Part 1–Geophysical survey.
34 Results of a gravity survey of the south-west margin of Dartmoor, Devon.
35 Geophysical investigation of chromite-bearing ultrabasic rocks in the Baltasound–Hagdale area, Unst, Shetland Islands.
36 An appraisal of the VLF ground resistivity technique as an aid to mineral exploration.
37 Compilation of stratabound mineralisation in the Scottish Caledonides.
38 Geophysical evidence for a concealed eastern extension of the Tanygrisiau microgranite and its possible relationship to mineralisation.
39 Copper-bearing intrusive rocks at Cairngarroch Bay, south-west Scotland.
40 Stratabound barium-zinc mineralisation in Dalradian schist near Aberfeldy, Scotland: Final report.
41 Metalliferous mineralisation near Lutton, Ivybridge, Devon.
42 Mineral exploration in the area around Culvennan Fell, Kirkcowan, south-western Scotland.
43 Disseminated copper-molybdenum mineralisation near Ballachulish, Highland Region.
44 Reconnaissance geochemical maps of parts of south Devon and Cornwall.
45 Mineral investigations near Bodmin, Cornwall. Part 2–New uranium, tin and copper occurrences in the Tremayne area of St Columb Major.
46 Gold mineralisation at the southern margin of the Loch Doon granitoid complex, south-west Scotland.
47 An airborne geophysical survey of the Whin Sill between Haltwhistle and Scots' Gap, south Northumberland.
48 Mineral investigations near Bodmin, Cornwall. Part 3–The Mulberry and Wheal Prosper area.
49 Seismic and gravity surveys over the concealed granite ridge at Bosworgy, Cornwall.
50 Geochemical drainage survey of central Argyll, Scotland.
51 A reconnaissance geochemical survey of Anglesey.
52 Miscellaneous investigations on mineralisation in sedimentary rocks.
52 Geochemical reconnaissance in the Cheshire Basin.
52 Titanium dioxide in the Ayrshire Bauxitic Clay.
52 The Marl Slate (Kupferschiefer) in the Southern North Sea Basin.
53 Investigation of polymetallic mineralisation in Lower Devonian volcanics near Alva, central Scotland.
54 Copper mineralisation near Middleton Tyas, North Yorkshire.
55 Mineral exploration in the area of the Fore Burn igneous complex, south-western Scotland.
56 Geophysical and geochemical investigations over the Long Rake, Haddon Fields, Derbyshire.
57 Mineral exploration in the Ravenstonedale area, Cumbria.
58 Investigations of small intrusions in Southern Scotland.
59 Stratabound arsenic and vein antimony mineralisation in Silurian greywackes at Glendinning, south Scotland.
60 Mineral investigations at Carrock Fell, Cumbria. Part 2–Geochemical investigations.
61 Mineral reconnaissance at the Highland Boundary with special reference to the Loch Lomond and Aberfoyle areas.
62 Mineral reconnaissance in the Northumberland Trough.
63 Exploration for volcanogenic sulphide mineralisation at Benglog, North Wales.
64 A mineral reconnaissance of the Dent-Ingleton area of the Askrigg Block, northern England.
65 Geophysical investigations in Swaledale, North Yorkshire.
66 Mineral reconnaissance surveys in the Craven Basin.
67 Baryte and copper mineralisation in the Renfrewshire Hills, central Scotland.
68 Polymetallic mineralisation in Carboniferous rocks at Hilderston, near Bathgate, central Scotland.

69 Base metal mineralisation associated with Ordovician shales in south-west Scotland.

70 Regional geochemical and geophysical surveys in the Berwyn Dome and adjacent areas, North Wales.

71 A regional geochemical soil investigation of the Carboniferous Limestone areas south of Kendal (south Cumbria and north Lancashire).

72 A geochemical drainage survey of the Preseli Hills, south-west Dyfed, Wales.

73 Platinum-group element mineralisation in the Unst ophiolite, Shetland.

74 A reconnaissance geochemical drainage survey of the Harlech Dome, North Wales.

75 Geophysical surveys in part of the Halkyn–Minera mining district, north-east Wales.

76 Disseminated molybdenum mineralisation in the Etive plutonic complex in the western Highlands of Scotland.

77 Follow-up mineral reconnaissance investigations in the Northumberland Trough.

78 Exploration for porphyry-style copper mineralisation near Llandeloy, south-west Dyfed.

79 Volcanogenic and exhalative mineralisation within Devonian rocks of the South Hams district of Devon.

80 Mineral investigations in the Ben Nevis and Ballachulish areas of the Scottish Highlands.

81 Investigations for tin around Wheal Reeth, Godolphin, Cornwall.

82 Mineral investigations near Bodmin, Cornwall. Part 4–Drilling at Royalton Farm.

83 Mineral investigations near Bodmin, Cornwall. Part 5–The Castle-an-Dinas wolfram lode.

84 An airborne geophysical survey of part of west Dyfed, South Wales, and some related ground surveys.

85 Geophysical surveys near Strontian, Highland Region.

86 Volcanogenic mineralisation in the Treffgarne area, south-west Dyfed, Wales.

87 Exploration for stratabound mineralisation in Middle Dalradian rocks near Huntly, Grampian Region, Scotland.

88 Mineral exploration for zinc, lead, zinc and baryte in Middle Dalradian rocks of the Glenshee area, Grampian Highlands.

89 Geochemical and geophysical investigations of the Permian (Littleham Mudstone) sediments of part of Devon.

90 Geochemical and geophysical investigations in Exmoor and the Brendon Hills.

91 A geochemical survey of part of the Cheviot Hills and investigations of drainage anomalies in the Kingsseat area.

92 A mineral reconnaissance survey of the Llandrindod Wells/Builth Wells Ordovician inlier, Powys.

93 Stratabound base-metal mineralisation in Dalradian rocks near Tyndrum, Scotland.

94 Geochemistry of some heavy mineral concentrates from the island of Arran.

95 Mineral reconnaissance at Menear, St Austell, Cornwall.

96 Geochemistry of sediments from the Lui drainage, Braemar, Grampian.

97 Magnetic and geochemical surveys in the area between Geltsdale, Cumbria, and Glendue Fell, Northumberland.

98 Exploration for gold between the lower valleys of the Erme and Avon in the South Hams district of Devon.

99 Base-metal and gold mineralisation in north-west Anglesey, North Wales.

100 Molybdenum mineralisation near Chapel of Garioch, Inverurie, Aberdeenshire.

101 Skarn-type copper mineralisation in the vicinity of Belstone Consols Mine, Okehampton, Devon.

102 Geophysical and geochemical investigations of the man ganese deposits of Rhiw, western Llyn, North Wales.

103 Exploration for volcanogenic mineralisation in the Devonian rocks north of Wadebridge, Cornwall.

104 Stratabound barium and base-metal mineralisation in Middle Dalradian metasediments near Braemar, Scotland.

105 Investigations at Lambriggan Mine, near St Agnes, Cornwall.

106 Marine deposits of chromite and olivine, Inner Hebrides of Scotland.

107 Mineral investigations near Bodmin, Cornwall. Part 6–The Belowda area.

108 Geochemical investigations around Trewalder, near Camelford, Cornwall.

109 Copper and molybdenum distribution at Shap, Cumbria.

110 Mineral investigations near Bodmin, Cornwall. Part 7–New uranium occurrences at Quoit and Higher Trenoweth.

111 Gold and platinum group elements in drainage between the River Erme and Plymouth Sound, South Devon.

112 Geophysical and geochemical investigations on Anglesey, North Wales.

113 Mineral investigations at Tredaule, near Launceston, Cornwall.

114 The Mineral Reconnaissance Programme 1990.

115 Platinum-group elements in the ultramafic rocks of the Upper Deveron Valley, near Huntly, Aberdeenshire.

116 Gold in the Ochil Hills, Scotland.

117 Exploration for vanadiferous magnetite and ilmenite in the Lizard Complex, Cornwall.

118 Mineral exploration in the Cockermouth area, Cumbria. Part 1–regional surveys.

119 Investigations for Cu-Ni and PGE in the Hill of Barra area, near Oldmeldrum, Aberdeenshire.

120 A gravity investigation of the Middleton Granite, near Inverurie, Aberdeenshire.

121 Exploration for gold in the South Hams district of Devon.

122 Mineral exploration in the Cockermouth area, Cumbria. Part 2: follow up surveys.

123 Mineral investigations in the Teign Valley, Devon. Part 2: base metals.

124 Platinum-group elements in the Huntly intrusion, Aberdeenshire, north-east Scotland.

125 Geochemistry database: data analysis and proposed design.

126 Mineral exploration in the Pitlochry to Glen Clova area, Tayside Region, Scotland.

127 The metalliferous mineral potential of the basic rocks of the Penmynydd Zone, south-east Anglesey.

128 Mineralisation in the Lower Palaeozoic rocks of south-west Cumbria. Part 1: regional surveys.

129 Mineralisation in the Middle Devonian volcanic belt and associated rocks of South Devon.

130 The occurrence and economic potential of nodular monazite in south-central Wales.

131 Platinum-group element mineralisation in the Loch Ailsh alkaline igneous complex, NW Scotland.

132 Reconnaissance drainage survey for base-metal mineralisation in the Lleyn peninsula, North Wales.

133 Exploration for gold in the Crediton Trough, Devon. Part 1: regional surveys.

134 Exploration for gold in the Crediton Trough, Devon. Part 2: detailed surveys.

135 The potential for diamonds in Britain.

136 A review of detailed airborne geophysical surveys in Great Britain.

137 Exploration for volcanogenic mineralisation in south-west Wales.

138 Gold exploration in the Duns area, Southern Uplands, Scotland.

139 Exploration for carbonate-hosted base-metal mineralisation near Ashbourne, Derbyshire.

140 Mineral exploration for gold and base-metals in the Lewisian and associated rocks of the Glenelg area, north-west Scotland.

141 Assessment of the potential for gold mineralisation in the Southern Uplands of Scotland using multiple geological, geophysical and geochemical datasets.

142 Industrial mineral potential of andalusite and garnet in the Scottish Highlands.

143 Gold mineralisation in the Dalradian rocks of Knapdale-Kintyre, south-west Highlands, Scotland.

144 The potential for gold mineralisation in the Britsh Permian and Triassic red-beds and their contacts with underlying rocks.

145 Exploration for stratabound mineralisation in the Argyll Group (Dalradian) of north-east Scotland.

146 Mineral exploration in Lewisian supracrustal and basic rocks of the Scottish Highlands and Islands.

MRP Data Release title

1 Data arising from investigations of mineralisation in Dalradian rocks, Tyndrum, Scotland.

2 Geochemical till sampling in the Middleton area, Peebles, Scotland.

3 Drill core from investigations around Chillaton, Devon.

4 Geochemical reconnaissance of the north-east Scottish Borders.

5 Data arising from investigations of the Lizard Complex, Cornwall.

6 Data arising from investigations in the Shelve area, Shropshire.

7 Data arising from drilling and overburden investigation at Brownstone Farm, Holbeton, South Hams, Devon.

8 Data arising from drilling investigations in the Loch Borralan intrusion, Sutherland, Scotland.

9 Data arising from investigations of the Knock intrusion at Claymires, Aberdeenshire.

10 Data arising from investigations into the distribution of platinum group elements, South Harris, Isle of Lewis, Scotland.

11 Rare earth elements in alkaline intrusions, north-west Scotland.

12 Mineral investigations in the Scardroy area, Highland Region, Scotland.

13 Exploration for gold in Central Wales.

14 Geochemical Surveys for gold in the Berwyn Hills.

15 An appraisal of the gold potential of mine dumps in the North Molton area, North Devon.

16 Exploration for stratabound mineralisation around Chillaton, Devon.

17 Regional appraisal of the potential for stratabound base-metal mineralisation in the Solway Basin.

18 Mineral investigations in the Northumberland Trough: Part 1, Arnton Fell area, Borders, Scotland.

19 Exploration for gold in the Thornhill Basin, Southern Scotland.

20 Mineral investigations in the Northumberland Trough: Part 2, Newcastleton area, Borders, Scotland.

21 Mineral investigations in the Northumberland Trough: Part 3, The Ecclefechan–Waterbeck area.

22 Mineral investigations in the Northumberland Trough: Part 4, the Bewcastle area.

23 Mineral investigations in the Northumberland Trough: Part 5, The Kirkbean area, south-west Scotland.

Appendix 3 Open file commercial mineral exploration projects carried out under the Mineral Exploration and Investment Grants Act 1972 (MEIGA)

Ref No AE	Project Name	Target Elements			Company
1	Wolfram Study	W			Consolidated Gold Fields Ltd
2	Scotland Copper/Nickel	Cu	Ni		Consolidated Gold Fields Ltd
3	Molybdenite Study	Mo	Cu	Ni	Consolidated Gold Fields Ltd
4	Loch Fyne and Cumlodden	Cu	Ni		Consolidated Gold Fields Ltd
5	Coed-y-Brenin	Cu	Mo		RioFinex
6	Ghrudie	Mo	W		Atlantic & Oceanic Resources
7	Brodiesord	Cu	Ni		Exploration Ventures Ltd
8	Huntly-Old	Cu	Ni		Exploration Ventures Ltd
9	Auchterless	Cu	Ni		Exploration Ventures Ltd
10	Boganclough	Cu	Ni		Exploration Ventures Ltd
11	Glass	Cu	Ni		Exploration Ventures Ltd
12	West Insch	Cu	Ni		Exploration Ventures Ltd
13	East Insch	Cu	Ni		Exploration Ventures Ltd
14	Cabrach	Cu	Ni		Exploration Ventures Ltd
15	Glenlivet	Mo	Cu	Ni	Exploration Ventures Ltd
16	Alford	Mo	W	Sn	Exploration Ventures Ltd
17	Kemnay	Mo	W	Sn	Exploration Ventures Ltd
18	Morven	Cu	Ni		Exploration Ventures Ltd
19	Lumphanan	Mo	W	Sn	Exploration Ventures Ltd
20	South Deeside	Mo	W	Sn	Exploration Ventures Ltd
21	Arthrath-Dudwick	Cu	Ni		Exploration Ventures Ltd
22	Belhelvie	Cu	Ni		Exploration Ventures Ltd
23	Kinharrachie	Cu	Ni		Exploration Ventures Ltd
24	Crichie	Cu	Ni		Exploration Ventures Ltd
25	Haddo	Cu	Ni		Exploration Ventures Ltd
26	Old Meldrum	Cu	Ni		Exploration Ventures Ltd
27	Udny	Cu	Ni		Exploration Ventures Ltd
28	Strichen-Crimond	Cu	Ni		Exploration Ventures Ltd
29	Newburgh	Cu	Ni		Exploration Ventures Ltd
30	Inverurie-Straloch	Cu	Ni		Exploration Ventures Ltd
31	Longside-Peterhead	Cu	Ni		Exploration Ventures Ltd
32	Slains	Cu	Ni		Exploration Ventures Ltd
33	West Haddo	Cu	Ni		Exploration Ventures Ltd
34	Maud	Cu	Ni		Exploration Ventures Ltd
35	Cruden	Cu	Ni		Exploration Ventures Ltd
36	Balquhindachy	Cu	Ni		Exploration Ventures Ltd
37	Quilquox	Cu	Ni		Exploration Ventures Ltd
38	Weardale	Pb	Zn	F	Acmin Explorations (UK) Ltd
41	Forest of Deer	Cu	Ni		Exploration Ventures Ltd
51	Sharnberry	Pb	Zn	F	SAMUK Ltd
52	Little Eggleshope	Pb	Zn	F	SAMUK Ltd
53	California	Pb	Zn	F	SAMUK Ltd
58	Clebrig	All	metallics		Acmin Explorations (UK) Ltd
59	Force Crag	Cu	Ba	Zn	Force Crag Mines
61	Durham Fluorspar	F	Pb	Zn	EXSUD Ltd
62	Parys-Mona	Cu	Pb	Zn	Noranda-Kerr Ltd
63	Rosehall	All	metallics		Oykel Minerals Ltd
64	St Ives Bay	Sn			Marine Mining Corporation
65	Godolphin Tin	Sn			Thyssen (GB) Ltd
66	FPA Continental	F			Continental Ore Ltd
67	Bonsall Fluorspar	F			EXSUD Ltd
69	Hudeshope	Pb	Zn	F	SAMUK Ltd
70	Harnisha Hill	Pb	Zn	F	SAMUK Ltd
72	Old Middlehope	Pb	Zn	F	SAMUK Ltd
73	Mulberry	Cu	Sn		Noranda-Kerr Ltd
74	Kilfinan-Glendaruel	Cu	Pb	Zn	Noranda-Kerr Ltd

75	Ffestiniog	Cu	Pb	Zn		Noranda-Kerr Ltd
76	Hafod-y-Llan	Cu	Pb	Zn		Noranda-Kerr Ltd
77	Prysor-Gamallt	Cu	Pb	Zn		Noranda-Kerr Ltd
78	Arenig	Cu	Pb	Zn		Noranda-Kerr Ltd
79	Pennant East	Cu	Pb	Zn		Noranda-Kerr Ltd
80	Pennant West	Cu	Pb	Zn		Noranda-Kerr Ltd
81	Rhiw	Cu	Ni			Noranda-Kerr Ltd
82	Ousdale	U				RioFinex
83	North Hucklow	F	Pb			Laporte Industries
84	Snitterton	F				Noranda-Kerr Ltd
85	Coulin	Cu	Pb	Zn		Noranda-Kerr Ltd
86	Hemerdon Mine	W	Sn			Hemerdon Mining and Smelting Ltd
87	Nantmor	Cu	Pb	Zn		Noranda-Kerr Ltd
88	Nantlle	Cu	Pb	Zn		Noranda-Kerr Ltd
89	South Kyle	All	metallics			Oykel Minerals Ltd
91	Craig	Cu	Ni			Noranda-Kerr Ltd
92	Dulyn	Pb	Zn			Noranda-Kerr Ltd
93	Drws-y-Coed	Cu	Pb	Zn	Ag	Kappa Explorations Ltd
95	Barra	Cu	Ni			Noranda-Kerr Ltd
96	Aberchirder	Cu	Ni			Noranda-Kerr Ltd
97	Brucklay-Maud	Cu	Ni			Noranda-Kerr Ltd
98	Tay-Comrie	Cu	Pb	Mo		Noranda-Kerr Ltd
100	North Molton	Cu				British Kynoch Metals Ltd
103	Ballantrae	Cu	Ni			Selection Trust Exploration Ltd
104	Glen Clova	Cu	Pb	Zn		Noranda-Kerr Ltd
105	Minera	Pb	Zn			Charter Consolidated Ltd
106	Elfordleigh	Sn	W			Consolidated Gold Fields Ltd
107	Huntly-New Bogie	Cu	Ni			Exploration Ventures Ltd
108	Cairnie	Cu	Ni			Exploration Ventures Ltd
109	Marnoch	Cu	Ni			Exploration Ventures Ltd
110	Ruthven	Cu	Ni			Exploration Ventures Ltd
111	Knock	Cu	Ni			Exploration Ventures Ltd
112	Closehouse	F				SAMUK Ltd
113	Assynt	Cu	Pb	Zn		Noranda-Kerr Ltd
114	Glen Calvie	Sn	Cu	Pb	Zn	Noranda-Kerr Ltd
115	Knapdale	Ni	Cu	Pb	Zn	Noranda-Kerr Ltd
117	Ramshaw	F				EXSUD Ltd
119	Duchy Peru	Cu	Pb	Zn		Texas Gulf Anglo Exploration Ltd
120	Cunningsburgh	Cu	Ni			RioFinex
123	Loch Awe	Cu	Ni	Zn		Consolidated Gold Fields Ltd
124	Deepwood	F	Ba	Pb		Acmin Explorations (UK) Ltd
125	Mill of Cairston	U	Pb	Ag	Zn	RioFinex
126	Dunlossit	Cu	Pb	Zn		RioFinex
127	Glengonnar Water	Pb	Zn	Cu	Au	RioFinex
129	Chasewater	Sn				Consolidated Gold Fields Ltd
130	Torpenhow	Cu				Consolidated Gold Fields Ltd
131	Nenthead	Pb	Zn	F		Mineral Industries Ltd
135	Knipes	Pb	Zn	Sb		SAMUK Ltd
136	Shetland	Cu	Ni			Noranda-Kerr Ltd
138	Loch Melfort	Cu				Noranda-Kerr Ltd
139	Slitt and Heights Veins	F	Ba	Pb		EXSUD Ltd
140	Molland	Cu				British Kynoch Metals Ltd
142	Mynydd-y-Garreg	Cu	Pb	Zn		Aquitaine Oil (UK) Ltd
148	Settlingstones	Ba	Pb	Zn		English China Clays Co Ltd
150	Vidlin	Cu	Zn			Grenmore Holdings Ltd
157	North Snailbeach	Pb	Zn			British Gypsum Ltd
158	Strontian Barite	Ba				Baroid UK Ltd
159	Castle-an-Dinas	Sn	W			South Crofty Ltd
160	Teign Valley	Pb	Zn	Ba		Black Rock Mineral Ventures Ltd
161	Killivose	Sn				Great Western Ores
162	Bedford United	Sn				South West Consolidated Minerals Ltd
163	Trenery's Prospect	Sn				Cornish Tin and Mining Ltd

164	Silver Valley	Sn				South West Consolidated Minerals Ltd
165	Redmoor	Sn	W			South West Consolidated Minerals Ltd
166	Rookhope Test Bores	F	Pb	Zn		SAMUK Ltd
167	Potash Drilling	K				Whitby Potash Ltd
173	Gairloch	Cu	Zn	Au		Consolidated Gold Fields Ltd
176	Joyce Vein/Gills Head	F				Wharfedale Mining
177	Killivose Tunnel	Sn				Great Western Ores
178	Restronguet Creek	Sn				Billiton UK Ltd
179	Witherite	Ba				Selection Trust
180	Killivose North Branch	Sn				Great Western Ores
182	Breney-Redmoor	Sn				Consolidated Gold Fields Ltd
184	Killivose DDH 136	Sn				Great Western Ores
185	Trewint & Tregirls	Sn				Hemerdon Mining and Smelting Ltd
186	Minera Lead/Zinc	Pb	Zn			Central Mining Finance Ltd
187	Hemerdon	W	Sn			Amax Hemerdon Ltd
188	Goonzion Downs	Sn				Geevor Tin Mines Ltd
189	Feasibility Study	K				Whitby Potash Ltd
190	Watertight Door 517	Sn				Great Western Ores
191	Reeves Lode	Sn				South Crofty Ltd
195	Dolcoath South Lode	Sn				South Crofty Ltd
196	Roskear Complex Lode	Sn				South Crofty Ltd
197	Red Vein Extension	F	Pb	Zn		SAMUK Ltd
198	Mulberry	Sn	W			Central Mining Finance Ltd
199	Devon Great Consols	Sn	Cu			Cominco (UK) Ltd
202	Allan's Shaft	Sn				Geevor Tin Mines Ltd
203	Dolcoath Branch and Mine	Sn				South Crofty Ltd
206	Arkengarthdale Dumps	Pb	Zn	F		Minex Minerals UK Ltd
207	Gallantry Bank	Ba				Baroid (UK) Ltd
208	Wheal Concord	Sn	W			Wheal Concord Ltd
210	Goss Moor	Sn				Billiton UK Ltd
211	Witheybrook Marsh	Sn				Geevor Tin Mines Ltd
212	Haye South	Sn	Ag	Pb		South West Consolidated Minerals Ltd
214	Whiddon Down	Cu	Pb	Zn		Amax Hemedon Ltd
217	Sunnyside Exploration	F	Pb			SAMUK Ltd
218	Gonamena	Sn	W			Black Rock Mineral Ventures Ltd
220	Nutberry Hill	Ba				RioFinex
222	Offshore Tin Alluvials	Sn				Billiton UK Ltd
223	Egloskerry	Cu	Pb	Zn	Ag	RioFinex
225	North St Austell	Sn	W			Billiton UK Ltd
226	Comrie	Cu	Pb	Zn	Mo	RioFinex
227	Witherite	Pb	Zn	Cu		BP Minerals plc
228	Great Wheal Carne	Sn				Geevor Tin Mines Ltd
229	Bothel	Cu	Pb	Zn		BP Minerals plc
230	Hopetoun	Cu	Pb	Zn		RioFinex
232	Tregullan	Sn	Cu	W		Central Mining Finance Ltd
233	Cairndow	Au	Cu	Pb	Zn	Cluff Ltd
235	Bridestowe	Cu	Pb	Zn		RioFinex
236	Scraithole	Pb	Zn	Ba	F	Black Rock Mineral Ventures Ltd
237	Fraddon Downs	Sn	W			Billiton UK Ltd
241	Sykes-Brennand	Pb	Zn			BP Minerals plc
243	Islay	Cu	Pb	Zn		Domego Resources
244	Clogau St Davids	Au	Ag			Caernarvon Mining
245	Treliver	Sn	W			Billiton UK Ltd
247	Great Flat Lode	Sn				Great Western Ores
248	Invergeldie	Cu	Mo	Au	Ag	RioFinex
249	Garths Ness	Cu	Pb	Zn		Grenmore Holdings
250	Cunningsburgh	Ni	Cr	Pt		Grenmore Holdings
251	Kelly Bray	Cu	Sn			South West Consolidated Minerals Ltd
253	Dalradian	Pb	Zn	Ba		Exxon
254	Sandlodge	Cu				Grenmore Holdings
256	Silver Hill	Sn				South West Consolidated Minerals Ltd
257	Hare Hill	Au				BP Minerals plc

258	Lagalochan	Au	BP Minerals plc
259	Moffat	Au	BP Minerals plc
260	Leadhills	Au	BP Minerals plc
261	Moorbrock	Au	BP Minerals plc
263	Lagolochan Extensions	Au	BP Minerals plc
264	Offshore Spectrometric	Sn	Geevor Tin Mines Ltd

Appendix 4 Selected BGS Radioactive and Metalliferous Minerals Unit Reports

Date	No.	Title	Authors
1964	257	Potash in the Fucoid Beds of Skye	Gallagher, M J.
1964	258	Fucoid Beds, Cambrian Formation, NW Highlands. Loch Broom to Kinlochewe, Ross & Cromarty	Dawson, J.
1964	259	Fucoid Beds, Cambrian Formation, NW Highlands. Kinlochewe to Loch Kisharr, Ross and Cromarty (summary)	Gallagher, M J.
1964	260	Field report on some measured sections through the Fucoid Beds in Sutherland	Dawson, J and Gallagher, M J.
1965	262	Re-examination of gold occurrences, Strath Kildonan, Sutherland, Scotland	Dawson, J.
1965	263	Proposals for uranium reconnaissance and research programme in the United Kingdom	Bowie, S H U.
1968	280	Reconnaissance mineral survey of King's Wood area, Buckfastleigh, Devon	Dawson, J.
1968	285	Preliminary report on heavy minerals in Northumbrian beach sands	Gallagher, M J.
1969	294	Helicopter-borne aeroradiometric survey of the cliffs of Caithness and Sutherland	Miller, J M and Rutter, S K.
1969	295	The radiometric anomaly at Allt na Muic, Helmsdale, Sutherland	Phillipson, A D, Greenwood, P and Gallagher, M J.
1969	296	Mineralisation in the Helmsdale granite and adjoining rocks, Sutherland. With notes on the petrology and geochemistry of the Helmsdale granite and its relationship to the Ousdale Arkose	Aitken, R N, Gallagher, M J, Phillipson, A D and Smith, R T, with Haslam, H W.
1970	303	Radiometric anomalies in the Helmsdale granite at Caen Burn, Sutherland, Scotland	Haynes, L.
1970	304	Uranium in the Old Red Sandstone of NW Caithness	Michie, U McL.
1970	305	Uranium mineralisation in the Ousdale Arkose, Caithness	Gallagher, M J.
1970	306	Radioactive occurrences at Houstry of Dunn, Caithness	Haynes, L.
1970	307	Reconnaissance for uranium in the Scottish Borders, 1969	Tandy, B C.
1970	308	Uranium and zinc in stream waters of Caithness	Haynes, L and Michie, U McL.
1970	311	Uranium mineralisation in the Ousdale Arkose - second report	Michie, U McL, Gallagher, M J and Simpson, A.
1971	312	Applied geochemical follow-up survey of the Ratagan area - Western Ross and Inverness-shire	Aucott, J W and Collingborn, C E M.
1971	313	Measurement of Cu, Pb, and Zn in United Uranium mill samples by portable radioisotope X-ray fluorescence analyser	Smith, R T.
1971	314	Gold in the Strath of Kildonan. Preliminary Report.	Plant, J A.
1973	315	A radiometric and geochemical reconnaissance of the Permian outcrop and adjacent areas in south-west England	Tandy, B C.
1973	319	Radiometric reconnaissance of north-east Scotland	Michie, U McL, Haynes, L and Cooper, D C.
1974	316	New radioactive nodule and reduction feature occurrences in the Littleham-Larkbeare area of Devon	Tandy, B C.
1974	321	Molybdenite mineralisation in Precambrian rocks, near Lairg, Scotland	Gallagher, M J and Smith, R T.
1974	322	Radiometric anomalies near St. Columb Major, Cornwall	Tandy, B C.
1974	323	Radiometric anomalies in the Tremayne area, near St. Columb Major,	Tandy, B C.
1974	324	Cornwall Radiometric and geochemical reconnaissance of the Fleet Granitoid, Kirkcudbrightshire	Tandy, B C.
1976	329	Investigation of the Larnrophyre intrusion at Shortlanesend, Cornwall	Cooper, D C.
1976	330	Investigations at Phernyssick, Cornwall	Cooper, D C.
1976	331	Investigations around St. Stephens, Cornwall	Cooper, D C.
1976	332	Geochemical till sampling at Vidlin, Shetland	Coats, J S.
1977	333	Geochemical soil survey at Sith Voe, Cunningsburgh	Aucott, J W.
1977	338	The distribution of U, Cu and Zn in the groundwaters of the Cheshire basin	Ball, T K.
1978	335	Bibliography of reports and publications on uranium in the United Kingdom	Michie, U McL L and Collingborn, C E M.
1979	336	Uranium concentrations of potential economic significance in Scotland and their genesis	Michie, U McL and Gallagher, M J.
1979	337	Uraniferous vein occurrences of south-west England - paragenesis and genesis	Ball,T K, Basham, I R and Michie, U McL.
1979	339	Uranium exploration in the English Midlands	Gunn, A G and Haslam, H W.

Appendix 5 High resolution airborne surveys in the UK for which BGS holds data

Survey Area	Client	Contractor	Year	Approx. Area (km^2)	Target	Line Spacing (m)	Terrain Clearance (m) Magnetics	EM	Radiometrics
1. Cornwall	UKAEA/BGS	Hunting G & G Ltd	1957	2,500	Mineral exploration	400	150	100	150
2 Devon & Somerset	UKAEA/BGS	Hunting G & G Ltd	1958/9	8,000	Mineral exploration	400	150	not flown	150
3 Argyllshire	BGS	Fairey Air Surveys	1961	1,500	Mineral exploration	300	not flown	not flown	150
4 Aberdeenshire	EVL	Barringer Research	1970	2,500	Mineral exploration	320	50	30	not flown
5 Anglesey	DTI	Hunting G & G Ltd	1972	230	Mineral exploration	100 & 200	30	30	60
6 Harlech Dome	DTI	Hunting G & G Ltd	1972/3	315	Mineral exploration	100 & 200	30 & 45	30	60
7 Bodmin	DTI	Hunting G & G Ltd	1973	60	Mineral exploration	200	45	30	60
8 Dent	DTI	Hunting G & G Ltd	1973	12	Mineral exploration	100	45	30	60
9 Augill	DTI	Hunting G & G Ltd	1973	6	Mineral exploration	100	45	30	60
10 Doon–Glenkens	DTI	Hunting G & G Ltd	1973	275	Mineral exploration	200	45	30	60
11 Lunedale	DTI	Hunting G & G Ltd	1973	50	Mineral exploration	200	45	30	60
12 Stockdale	DTI	Hunting G & G Ltd	1973	100	Mineral exploration	200	45	30	60
13 Craven	DTI	Hunting G & G Ltd	1973	175	Mineral exploration	200	45	30	60
14 Lothersdale	DTI	Hunting G & G Ltd	1973	16	Mineral exploration	200	45	30	60
15 Blair Atholl	DTI	Sander Geophysics Ltd	1974	325	Mineral exploration	200	35	45	60
16 S Northumberland	DTI	Sander Geophysics Ltd	1978	440	Mineral exploration	250	60	75	75
17 Girvan–Ballantrae	DTI	Sander Geophysics Ltd	1978	365	Mineral exploration	250	60	75	75
18 S W Dyfed	DTI	Sander Geophysics Ltd	1978	670	Mineral exploration	250	60	75	75
19 Dalradian	EXXON	Dighem Ltd	1983	430	Mineral exploration	200	50	35	not flown
20 Witham & Tiptree	BGS	Barringer Research	1973	160	Sand and Gravel	160	not flown	60	not flown
21 Garstang	WRB	Barringer Research	1973	120	Hydrogeology	320	not flown	60	not flown
22 Formby	BGS	Global Earth Sciences	1987	100	Hydrocarbons	200	100	not flown	not flown
23 Fylde	BGS	Global Earth Sciences	1987	100	Hydrocarbons	200	100	not flown	not flown
24 Sellafield	NIREX	Aerodat ltd	1990/91	500	Waste Disposal	200	100	115	130
25 Dounraey	NIREX	Global Earth Sciences	1991	400	Waste Disposal	200	100	not flown	130
26 Midlands HIRES	BGS	World Geoscience	1998	14,000	Regional Survey	400	90	90	radiometric were flown